The Role
of Environmental Impact Assessment
in the Planning Process

The Role of Environmental Impact Assessment in the Planning Process

edited by
MICHAEL CLARK and JOHN HERINGTON

An Alexandrine Press Book

MANSELL PUBLISHING LIMITED
London and New York

First published 1988 by
Mansell Publishing Limited, *A Cassell Imprint*
Artillery House, Artillery Row, London SW1P 1RT, England
125 East 23rd Street, Suite 300, New York 10010, U.S.A.

This book was commissioned, edited and designed by
Alexandrine Press, Oxford

British Library Cataloguing in Publication Data
The role of environmental impact assessment in the planning process.
  1. Great Britain. Environment planning.
  Environmental impact analysis
  I. Clark, Michael   II. Herington, John
  333.7'1'0941

  ISBN 0-7201-1889-1

Library of Congress Cataloging in Publication Data
The Role of environmental impact assessment in the planning process /
  edited by Michael Clark and John Herington.
      p.   cm.
    "An Alexandrine Press book."
    Includes index.
    ISBN 0-7201-1889-1 : $60.00 (U.S. : est.)
    1. Environmental impact analysis——Great Britain.
  2. Environmental impact analysis——European Economic
  Community countries. 3. Environmental protection——
  Great Britain——Planning. 4. Environmental protection——
  European Economic Community countries—Planning. I. Clark,
  Michael, 1949-   . II. Herington, John.
  TD194.6.R85   1988
  333.7'15——dc19
                                                      88-12259
                                                          CIP

This book has been printed and bound in
Great Britain: typeset in Paladium by
Colset (Private) Ltd., Singapore, printed on
Redwood Book Wove paper by Redwood
Burn Ltd., and bound by WBC Bookbinders.

Contents

Preface

The idea for this book sprang from a conference organized by the Geography and Planning Study Group of the Institute of British Geographers and held at Loughborough University in 1979 (Herington, 1979). The papers produced then still seem to us today to be surprisingly fresh and to provide an excellent starting point for up-dating developments in EIA during the 1980s. Admittedly the subject of Environmental Impact Assessment has become a well-trodden field and other interesting seminars have followed (see, for instance, Breakall and Glasson, Environmental Impact Assessment: From Theory to Practice. Oxford Polytechnic Department of Town Planning, 1981), but in our view they have substantially reproduced the ground covered at the original Loughborough meeting.

We initially conceived the book as a series of case studies which would give a retrospective evaluation of past practice and help to remind government of the advantages of introducing a more formalized approach to EIA in the public planning system. At the time this seemed a wise approach, bearing in mind the government's tardy acceptance of the principle of environmental controls over industry. What we had not anticipated was the British government's acceptance of the EC Directive. The new procedures and approaches being introduced in Britain in 1988 pose considerable problems and uncertainties for the business community, government and the growing environmental lobby. Before the EC Directive comes fully into effect it is important to take stock of past practice among those who have experience of working with EIA – bearing in mind that the wider social and political implications of EIA will remain important well beyond this date. We therefore wanted the book to integrate the experience of EIA with a forward look at the scope, opportunities and constraints presented by the advent of new legislation.

Many books already emphasize the specialized methods needed to undertake EIAs. Another book on these lines seems inappropriate at this time. More significant, in our view, is the need to examine the uses and implications of EIA in relation to the political processes of decision-making and planning. Our interest is less in how EIA may be carried out and rather with why and to what effect EIA has been used in different situations – the extent to which the procedure has proved itself beneficial and if not, why not. We seek a wider-ranging view of the applications and value of EIA and hope to provide some answers to the question: has EIA been a useful tool for planning or not?

The book is written as much for the general reader as the student of EIA methodology. Our intention is to provide an introduction for those unfamiliar with the practical experience of EIA in Britain, or seeking a knowledge of the structure of the new EC Directive and its ramifications for the public planning system. We would expect the book to have both general and specialist interest for government agencies, practising planners and developers in the public and private sector. We hope there is something of interest for the environmental movement, particularly those groups involved in formulating and challenging development policy.

Bringing together the original authors has not been without its problems. We must thank all those contributors who have helped to meet deadlines. Over such a long period many people, especially the colleagues in our respective departments, have been helpful. Finally, we must thank both our families for their patience and support.

Michael Clark and John Herington
Preston and Loughborough, 1988

The Contributors

A.Bell, BA, Principal Planning Officer, Cheshire County Council.

I.Brown, BA. PhD, Principal Planning Officer (Environment), Department of Architecture, Planning and Estates, Clwyd County Council.

M.Clark, BA, MSc, PhD, Senior Lecturer in Geography, School of Construction, Surveying and Geography, Lancashire Polytechnic.

D.R.Cope, BA, MSc. International Energy Agency, Coal Research.

J.M.Herington, MA, MPhiL, MRTPI, Lecturer in Planning Studies, Department of Geography, Loughborough University.

P.Hills, BA, MA, PhD, Institute of Planning Studies, University of Nottingham.

G.McDonic, Dip TP, FRTPI, DPA, Barrister at Law, County Planning Officer, Wiltshire County Council; President of the Royal Town Planning Institute, 1985–86.

D.Roberts, BA, PhD, Research Assistant, Department of Architecture, Planning and Estates, Clwyd County Council.

R.G.H.Turnbull, Deputy Chief Planner (now retired), Scottish Development Department.

P.Wathern, BSc, PhD, Lecturer in Botany and Microbiology, University College of Wales, Aberystwyth.

K.Williams, MA, PhD, Lecturer in Regional Planning, Institute of Planning Studies, University of Nottingham.

R.H.Williams, BA, McD, MRTPI, Lecturer in Town Planning, Department of Town and Country Planning, University of Newcastle Upon Tyne.

C.Wood, BSc, DipTP, MA, MRTPI, Senior Lecturer in Town and Country Planning, University of Manchester.

S.Young, BSc, PhD, MIWPC, Department of Botany and Microbiology, University College of Wales, Aberystwyth.

Introduction: Environmental Issues, Planning and the Political Process

MICHAEL CLARK and JOHN HERINGTON

The extent to which the developed and developing worlds have taken up the philosophies and practice of environmental impact assessment is remarkable considering the cultural, social and political differences between nations. The sheer scale of infrastructure projects as well as the continued evidence of man's unthinking approach to resource exploitation have helped to encourage greater awareness of environmental issues. Threats to human health, food and fuel supplies, and the long-term consequences for the biosphere are questions of major importance to all of us who share concern about the future of man on this planet (Strong, 1984).

In many countries environmental issues are subject to growing political debate. The economic case for development must be weighed alongside environmental factors. Organized pressure groups such as Greenpeace are increasingly effective in questioning the ethics of unbridled national government practice and the development role of big companies. The public demand for open debate of controversial environmental issues has made life more difficult for decision takers: it has forced government to articulate public interest in environmental protection and social well-being more clearly, even to the extent that developments which might have seemed inevitable have sometimes had to be temporarily shelved, as was the case with the test drilling programme in Britain. More sophisticated legal and financial controls are introduced by governments to regulate the actions of private industry; and delegated responsibilities for pollution licensing and environmental monitoring are coming under new national and international controls (i.e. the new inspectorate for water quality).

The common response of governments to these changing conditions has been to reform existing legislative procedures to permit more explicit treatment of environmental factors in decision making. Such approaches secure approval from those on the Right of the political spectrum who argue that although environment is important nothing fundamental must be done to prevent future economic growth – indeed economic development is necessary if environmental improvements are to follow. In contrast, these reformist approaches to environmental problems displease those Liberals or Social Democrats who question the direction of economic policy and wish to see introduced a much higher degree of public accountability on the part of industry and government; and for Marxists nothing less than the control of capital is likely to satisfy their aspirations for the environment. All these positions may be criticized as lacking the minimum requirements for a 'green' ideology, which is fundamentally different in analysis and values.

The scope of radical measures to alleviate the worst effects of man's impact on the environment is severely limited by economic, social and political factors. Here there is a dichotomy between those who make a virtue of technical and political reformism and those who believe that only by adopting entirely new environmental priorities can government and business ensure harmony between man and nature. The first view is well put by O'Riordan and Turner (1983, pp.10) who argue that:

> the real message of environmentalism lies . . . in the rather vague areas of analytical technique, political process and ecocentric understanding . . . the enlightened technocentrist of tomorrow ought to be the true environmentalist – the environmental manager who seeks to reduce the adverse consequences of economic development and social improvement to acceptable levels.

The difficulty with this kind of environmentalism is that it implies an inherently conservative attitude to the social and economic policies which bring about development. If the views of the environmental manager accord with those of the politician a fusion of economic and environmental goals might be achieved – if they do not, there is no necessary reason why environment values should triumph. A further difficulty with the O'Riordan argument is that it places great faith in the far-sightedness of the environmental manager, a particular problem when the pressures for economic development are strong.

An alternative scenario puts faith in political solutions to environmental problems. As Grove-White argues (in Roberts and Roberts, 1984, p.148)

> Does not the need for Environmental Impact Assessment arise, in this country as in the US, from a recognition of a quite new order of problems for our political systems – from conflicts between man and nature, between the escalating demands of industrial society and the limits of the biosphere and its resources? If so, won't the conception of EIA as a checklist of impacts at project level make only a tiny contribution to addressing these problems?

This scepticism of procedural reforms is perhaps inevitable given the enormity of environmental problems. The best chance for convergence of opinion will be found in the more widespread adoption of environmental values in different countries of the world. If more radical policy stances are to emerge from governments they may do so only as the result of a broadening of the definition of 'environmentalism', a much greater injection of environmental values in public and private life, and the emergence of environmentalist political candidates (Slocumbe, 1984). In Britain, as in other developed countries, the main political parties show growing concern for the 'environmental vote' and there is evidence that public pressure will temper the more extreme views of government. However, there may also be a growing gap between those at work who share environmental concerns and those out of work who reject them.

Environmental Impact Assessment

Environmental Impact Assessment (EIA) is a structured approach and formal set of procedures for ensuring that environmental factors are taken into account at all levels in planning decision taking. It is fast developing in Britain and other countries of the world in response to public concern about the health, safety, amenity and social consequences of major investment projects such as nuclear power stations, coal mines, oil installations and transportation routes, but as yet has been very limited in its application to national government policies which affect the environment. EIA is a concept which has become politically sensitive especially when proposals for mandatory systems of environmental control have been introduced. However, most commentators view EIA essentially as a reformist tool to aid planning decision taking rather than an environmental protection measure *per se*.

> There is common agreement that the fundamental aim of EIA is not to determine the balance placed by the decision-maker on environmental compared to economic, social or other considerations but to ensure that the decision is made on the basis of informed knowledge of the environmental consequences of that decision. (Roberts and Roberts, 1984, p.100)

> EIA should be thought of as a concept which allows decision-makers to be made aware of likely impacts should a proposal proceed. Whether or not it does proceed may be influenced by other factors such as energy needs, employment levels or technological necessity, which legitimately may have greater priority at any given place or time. (Clark and Turnbull in Roberts and Roberts, 1984, p.136)

> Environmental Impact Assessment is not an environmental protection measure . . . the requirement for an EIA suggests that some thought is being given to environmental consequences . . . and feeds awareness of the need for balanced development and sustainable environment. (Bidwell, 1985)

The literature on EIA techniques, methods and procedures has proliferated during the past fifteen years. The history of EIA may be seen as a progression from the limited obligation placed upon the proponents of an action to make specific an assessment of the environmental consequences of their proposal, as under the United States National Environmental Policy Act (NEPA) 1969, towards a more wide-ranging identification and evaluation of a large variety of consequences arising out of public or private decisions. Yet it would be dangerous to assume that there is anything natural or correct in this extension of the uses to which EIA is put, or of the degree of political authority or bureaucratic sophistication necessary to enable it to have effect. In the future EIA could be used either as a means for governments to impose greater controls and safeguards on developments and innovations of all kinds or, alternatively, as a vehicle for reducing the power and influence of political groups intent on objecting to any new actions which will bring about change in the environment.

There is no universal agreement as to the subject matter of EIA, although four broadly distinct, though not mutually exclusive, spheres of impacts have been identified: ecological impacts, social impacts, technology impacts and risk or hazard impacts (see Cope and Hills, Chapter 10 in this volume for a full discussion). Despite a desire to see EIA used to identify *all* the range of impacts arising from an action, in practice studies have tended to include those impacts for which some measure of quantified information could be obtained, such as land-use change and habitat displacement, soil, water and air pollution, climatic factors, species variety, or noise levels. But the environment includes people and their values which increasingly shape public responses to project development. In America there has been a lively interest in that branch of assessment known as 'Social Impact Assessment' (SIA). In contrast, European government agencies have tended to play down the significance of non-ecological factors such as migration or cultural values. The absence of concern for social assessment is clear from an inspection of those EIAs carried out in Britain (Petts and Hills, 1982), though it is accepted there is a wide measure of disagreement as to what constitutes a 'social' impact.

There is little doubting the importance of EIA as a concept. Where formal systems exist, a project's success depends on how well potential developers can demonstrate that the effects on the environment are not harmful to society. The central question is how EIA is made to work within the established political and institutional frameworks in different countries and, in turn, what effect these frameworks will have upon the effective implementation of EIA.

Australia, Canada, France, Germany, Ireland, Japan, New Zealand, Spain and the United States have already implemented formalized EIA systems. On June 27, 1985, the European Communities (EC) published a Directive on 'the assessment of the effects of certain public and private projects on the environment' which identifies in two lists (Annex 1 and 2) those projects which by virtue of their nature, size or location are likely to have significant effects on the environment and thereby require EIA. All the Member States of the European Economic Community are required to adhere to the Directive although the form and methods of implementation are left to the national authorities. The British government, after a period of initial uncertainty, has agreed to the terms of the Directive and will publish legislation before July 1988 indicating how EIA is to be incorporated within the British planning system.

In Britain the debate about whether EIA should be incorporated into the existing planning system is thus effectively ended. The Department of Environment has published a consultation paper giving guidelines on which Annex II projects may be candidates for EIA and the criteria, thresholds and procedures to be used (DoE, 1988). Yet there remain many uncertainties about the future use of EIA in Britain. What range of developments will be subject to impact assessment? Which projects will be judged to have 'significant' environmental effects and thereby require environmental assessment? To what extent and at what stage will the public be involved? Could a local planning authority refuse a planning application because it was dissatisfied with the environmental assessment carried out by the developer? Would the Department of Environment uphold a planning decision based on these reasons? How much weight will be given by Inspectors to the evidence of EIAs at public inquiries? Has local government the necessary technical competence and expertise to carry out EIA? Perhaps most importantly, will the existence of EIA help bring about a climate of public opinion which would favour strengthening planning controls over agriculture and forestry, both of which activities are currently exempted from planning control?

Although the actual number of developments which are treated as 'significant' may be rather few in the early years of introducing EIA into the planning system, most commentators agree that the imminent prospect of EIA has considerable implications for future planning practice in Britain, for the quality of environment and, possibly most important, for the way in which land development decisions are arrived at in a democracy. The introduction of EIA could emphasize the importance of greater environmental accountability both to industry and to Departments of State. Whether EIA by itself is capable of changing public expectations about the environment is more doubtful. The politics of the green movement seems likely to be more influential than EIA in achieving a radical shift of opinion. In this context, EIA may prove to be something rather incidental, peripheral and perhaps conservative – even if it is a step forward.

We should avoid too much euphoria over the prospect of mandatory EIA. The political, economic and institutional context within which development projects are formulated and will be assessed has undergone profound change since EIA techniques were first being talked about in Britain. In part this is because new kinds of environmental issue have surfaced and in part because the national economic and political climate has changed, bringing with it a questioning of the virtues of public controls over the environment at a time of economic recession. One important consequence has been the gradual erosion of the public planning system into which EIA will be incorporated. All these issues affect our judgement about the likely future effectiveness of EIA and require comment.

The Politics of Environment in the 1980s

Environmental issues are increasingly international and national in scale (Strong, 1984). The most pressing problems facing governments are nuclear policy, acid rain and the transformation of the countryside. Happily there are at last signs that politicians are becoming aware of the 'greening' of public opinion in Britain. The 'ecology versus jobs debate' is on the political agenda, not least because of the efforts of pressure groups such as the Friends of the Earth.

The global implications of nuclear proliferation and the dumping of radio-active nuclear waste (Cope, 1984: Heath, 1985) have dramatically aroused public and political opinion in the wake of the Chernobyl disaster and anxieties about radioactive leaks at Sellafield. The case for additional nuclear power stations is being reassessed by all the political parties, though with little radical change in position obvious. Labour reluctantly concede to the TUC view that nuclear should not be scrapped. The Alliance favours a balanced energy strategy which includes nuclear power. The Conservatives continue to be pro-nuclear. The anti-nuclear position is maintained most strongly by the Green Party. The extent to which EIA can help evaluate nuclear issues remains proble-matic. The EIA of the proposed European fast reactor reprocessing plant planned for Dounreay in northern Scotland has been described as a 'peripheral document rather than one of real analysis and utility' (Ryder, Bisset and Kayes, 1986). Inadequate consideration of risk and socio-economic assessment is reported.

The impact of acid rain on American and European forests has been a continued source of concern. Britain has finally agreed to spend £600 million on fitting desulphurization equipment to three power stations, although this is not sufficient to allay government and conservationist interests in Germany and Scandinavia (Rose, 1984). Acid rain is another example of the international dimension to environmental issues in the 1980s, although atmospheric pollution has proved to be an emotive political issue at the local scale (see, for example, Blowers, 1984).

Environmental issues in the British countryside arouse increasing political concern, especially the social, economic and visual changes affecting the land-scape. The clashes of interest between developers and conservationists, which Gregory described so well in his book *The Price of Amenity* (1971), are still present, although new kinds of conflict are apparent. With evidence of mounting agricultural farm surpluses the debate is more about the future use of farmland that may no longer be required for food production than about the loss of farmland to urban development – although the dangers of an unintended and gradual urbanization of the countryside have not receded (see, for instance, the Council for the Protection of Rural England, 1981). Unusually rapid political commitment to proposals for large-scale transport projects, such as the Channel Tunnel, Stansted Airport and the M25, has raised major environ-mental and planning issues. The North Sea oil boom has come to an end, although the search for new oil supplies resurfaces in scenically attractive parts of southern Britain and in the Irish sea. The role of Environmental Impact Assessment in helping to formulate these policies and projects remains limited and uncertain.

The future of the Green Belt is a thorny issue for the present government given its insistence on the need to stimulate economic recovery, although the pressures for incursion are greatest in London and the South East (Short, 1984). Old fears have been rekindled about the visual impact of single species afforestation in the uplands (Nature Conservancy Council, 1986; Countryside Commission, 1986). Landscape planning in the National Parks remains a central issue (MacEwen A. and MacEwen M., 1982; Anderson, 1985). The protection of wildlife and natural habitats was highlighted in 1985 by the conflict on Islay between the Nature Conservancy Council and the Scottish Malt Distillers (Stroud, 1985). Commentators have drawn attention to the rising loss of Sites of Special Scientific Interest (Goode, 1984; Mills, 1986) and the loopholes in the Wildlife and Countryside Act, 1981.

The impact of modern farming methods on the appearance of the countryside (Shoard, 1980) has become an important issue for the political parties. The need to control agrarian change formed part of the 1986 Labour Party's environmental programme. Adjustment of subsidies from automatic price support into schemes to protect the appearance of the countryside and help poorer farmers are proposed. Moreover, under this programme farmers would have to pay rates on agricultural land and seek planning permission for all farm buildings and development including the removal of hedgerows; the National Farmers Union have predictably declared their opposition to both these reforms. The Conservatives have agreed to pay farmers to protect conservation interests in twelve designated areas of countryside in England and Wales. These are labelled 'Environmentally Sensitive Areas' (ESAs). Definition of an ESA depends in part on whether the landscape is threatened by farming practices. The Countryside Commission proposed an initial short-list of areas from which the final selection was made. The areas include the Norfolk Broads, the Pennine Dales, the Somerset Levels and Moors, the eastern half of the South Downs, west Penwith and parts of the Cumbrian mountains. A total of £12m is paid annually to farmers to encourage them to continue traditional farming methods. The concept has been criticized by farmers but welcomed by the amenity organizations.

How can EIA respond to the changing politics of the environment? The concept and methods of EIA were introduced in the early 1970s to deal with the impact of large-scale oil installations. The nature of the environmental debate has shifted considerably since then. Despite advocates of a more extensive use of EIA in policy issues, EIA has remained at the project level for the most part. Major proposals will still generate local political debate and a more widespread application of EIA could help to air all the issues in a dispute. But the real challenge lies in relating procedures to the increasing scale of environmental concern over such issues as nuclear dumping, the urbanization of the countryside and environmentally damaging agricultural development. It is difficult to see how this can happen while EIA is largely ignored at the point of political and policy debate.

New Political Goals for Planning

The political context in the mid-1980s is radically different from that prevailing when impact assessment procedures were first being introduced. The policy of turning over State-owned enterprises to private ownership ('privatization') has been pursued in the water supply industry, though shelved in mid-1986, and is being proposed for the forestry industry. Both raise awkward questions about the possible future quality and effectiveness of EIA carried out by private bodies.

The ideology of the Thatcher government stresses the enabling element of planning, a framework for action as opposed to a system of negative controls.

> Planners must help create the right conditions and ensure that business initiatives prosper . . . what does concern me is that planning procedures should not hamper the economic recovery now slowly emerging . . . Planning authorities must adopt a flexible and pragmatic approach to meet the needs of versatile enterprises. Don't tie them down to conditions or restrictions that were designed for the more traditional smokestack industries. (Text of an address to the Royal Town Planning Institute by the Secretary of State for the Environment, Patrick Jenkin, MP, 1984)

There is a straightforward case for introducing EIA into a remodelled planning system which is geared primarily to improving the efficiency of the business sector. EIA may be used as a means of overcoming political or localized barriers to change and thus be within the spirit of an enabling philosophy. For instance, while the EC Directive on EIA has been accepted in principle by the British government, the wording of the Department of Environment's initial consultation document shows that the argument of the Confederation of British Industry that environmental legislation must not inhibit or delay necessary industrial projects has been accepted. The counter proposition is that environmental impact techniques produce savings to industry in the time taken to achieve planning consents, especially if EIA is brought into decision making at the earliest possible stage (Dean and Graham, 1977).

Financial savings are likely in practice only when significant public opposition to development proposals has been avoided. Not opening EIAs to public inspection, or at least delaying the point when the public will be brought in, would seem an attractive option for those who seek *less* government and who want rational, objective or scientific mechanisms to avoid the delay and timid decision making associated with established political procedure – particularly when vociferous amenity lobbies and other interest groups become involved. Yet if the public voice is effectively stifled and the planning system is geared more and more to the goals of economic development it would seem difficult for EIA to be used as a neutral tool in decision making, a theme Herington returns to in Chapter 9.

Since the early 1970s there has been a perceptible shift in public values towards natural resources and the natural environment – witness the growing importance of the 'green' vote in politics (Porritt, 1984 p.10; Rees, 1985, p.378). If more politicians come to share the values of the 'greens' the current ideology in public planning could change direction. We should not forget that the early purposes of the town and country planning movement were born from a belief that left unfettered, markets in land development would not secure specific environmental values. The political goals for planning today are different. Fitting EIA into a planning system with explicitly stated environmental objectives might seem at best simplistic and at worst unnecessary. The British government initially opposed the introduction of EIA. Some politicians still suspect that good environment and strong economy are seen to be mutually exclusive – although this view is not shared by other European countries.

That the government's view is too narrow is shown by the emergence of a new environmental economics arising from the basis of better regard for the costs of utilizing finite natural assets and of a more realistic appraisal of the full costs of negative externalities as well as the justice of seeing that such externalities are borne by the user rather than society as a whole (Fairclough, 1984). Environmental protection will therefore always prove worthwhile for industry wherever the risks, as expressed by the costs of damage or danger, exceed the costs of control, management or amelioration. This links with Hardin's analysis of the hazards facing an 'open access' resource:

> The tragedy of the commons as a food basket is averted by private property, or something formally like it. But the air and waters surrounding us cannot readily be fenced, and so the tragedy of the commons as a cesspool must be prevented by different means, by coercive laws or taxing devices that make it cheaper for the polluter to treat his polluters than to discharge them untreated. (Hardin, 1968, p.79)

This is the 'polluter pays' principle which underlies the European Communities' environment programme. However, Hardin argues that the law is generally so behind the times that its sanctions tend to be ineffective. In reality, our shared environment is likely to deteriorate until it threatens not just our amenity but our very existence. Like Malthus, Hardin blames population growth. His 'lifeboat ethic' prescribes a drastic attack on fertility in the Third World (Pepper, 1984, p.209). Such 'ecofascism' has been criticized on moral and pragmatic grounds, and it can be argued that his assumption about a lack of social and economic controls over the use to which 'commons' are put is wrong (Rees, 1985, p.253). None of this detracts from widespread contemporary anxiety about environmental issues and dissatisfaction with established thinking and practice, especially in the narrow specialism of economics. The work of The Other Economic Summit promises an important new perspective and set of priorities:

a New Economics, based on personal development and social justice, the satis-
faction of the whole range of human needs, sustainable use of resources and
conservation of the environment. (Ekins, 1986, p.xv)

Until such an interpretation is adopted, there will remain a need for intervention or control. Here EIA may also be seen as a way of positively intervening
in the market economy to achieve direct bureaucratic or political means to ideological ends. In this context EIA can help governments impose controls and safeguards on developments and innovations of all kinds. This philosophy is seen in
all centrally planned economies, many Socialist policies and evident in aspects
of the FAST programme of the EC (Commission of the European Communities,
1984) and also the European Commission for Europe (ECE).

Environmental Impact Assessment cannot of itself secure the safeguarding of
environmental values. However, we do not accept that EIA can be a value-free
scientific method. In practice the scope, methods and utility of impact assessments will be influenced by prevailing political goals. The question is: what
goals are most appropriate? The rationale for using EIA under NEPA was the
underlying belief that environmental values were being given too little weight in
decision making. We would argue that there is little purpose in devising new and
improved techniques if that premise no longer holds.

Questions for the Future

The operational context for EIA – or EA, if one accepts the official British
shorthand – will be a small number of major, large-scale, potentially dangerous
or otherwise significant or contentious projects. Which projects will be covered
will depend at first on how European guidelines and requirements are interpreted. It is not expected that EIA will be applied to policies or plans or programmes in the near future. It will mainly concern big schemes of the sort which
have generally been subject to quasi-judicial scrutiny at public inquiries, and in
some cases been the topic of Parliamentary debate.

It will be important to evaluate EIA's contribution as a formal addition to the
UK public planning system at all stages from formulation of development plans
to decisions about specific projects, including the public inquiry system.
Experience already gained with informal and non-statutory EIAs used in the
strategic planning process has been limited although its scope is considerable, as
demonstrated in the first chapters of this book. Where EIA has been used as part
of the evidence presented by the developers at public inquiries the evidence from
Williams's research, outlined in Chapter 4, is that their credibility will rest on
the extent to which impact assessments have included factors beyond the
control of the developer. In some cases an EIA will improve the applicant's case.
More options are kept open and the tendency for absolute commitment to a
single course of action is restrained. Scrutiny at the inquiry is more effective, as
EIA can improve the quality and amount of information available. As a result
Ministers receive better quality advice, and may be able to make more readily

accepted (or 'legitimized') decisions. But the nagging question remains: how much political commitment will there be to the spirit as well as the letter of EIA?

The public planning system is in disarray in the mid 1980s. Some commentators have drawn attention to the gradual death of strategic thinking, attributing this to the desire 'to get things done' (Breheny and Hall, 1985). The Nuffield Commission concludes its investigation of the planning system with a plea for stronger national policy guidelines (Nuffield Foundation, 1986). A Royal Town Planning Institute Study Group has argued the case for a complete rethink of current land policies within a new regional strategic framework (Royal Town Planning Institute, 1986). There is mounting concern about the delays and inequities posed by the public appeal system (House of Commons Environment Committee, 1986), a matter over which the Town and Country Planning Association has lobbied government for some time.

Those seeking greater control over the environmental impact of forestry and farming operations argue strongly that the planning legislation should be amended to include these as 'development' within the meaning of the Town and Country Planning Acts. Yet the political battles for supremacy among government Departments of State seem likely to stifle any more rational distribution of statutory powers over the environment. Conflict between the Ministry of Agriculture, Forestry and Fisheries and the Department of Environment over the conservation of rural resources remains unresolved, with some arguing that the DOE should assume greater responsibility for farmland controls or that a new Ministry of Environmental Protection is required which would take on board the countryside functions of MAFF.

Of more general interest is whether existing planning controls and procedures are sufficiently democratic to protect ordinary citizens from the adverse environmental and social effects of major new development. When EIA was being tested by the Department of Environment's consultants in the mid-1970s the development control and development plan system had not been subject to the rigorous modification they experienced in the first half of the 1980s. The legislative changes in planning introduced by the Local Government, Planning and Land Act, 1980, have had the effect of weakening strategic (regional and county) planning control at a time when many issues of public concern have implications beyond the boundaries of individual local authorities boundaries. The relaxation of normal planning controls in Enterprise Zones and the advent of Simplified Planning Zones have reduced the occasion for public debate about the environmental implications of major new developments, as well demonstrated by the case of the Canary Wharf scheme for large-scale speculative office development in London's docklands (for a review of recent change in the planning system, see Greater London Council, 1985).

Changes in planning legislation have been brought about by the present government's concern for economic recovery and the removal of the burden of legislation which might restrict the freedoms of enterprise to locate where it wishes. The old argument that Britain already has a working and effective system of environmental impact assessment, namely the development control procedures, looks a little thin at a time when normal planning requirements are less likely to be refused because of changed government policy. Moreover, the subtle changes being made to planning procedures influence how well environmental assessment works as an adjunct to existing development planning and control: the more limited extent of public participation, the reduced scope for research, the requirement on planning officers to shorten the time taken in reaching decisions, are all matters which increase the danger of short-term thinking and may limit the potential utility of EIA in the planning process.

The Structure of the Book

Much has been written on EIA as our references indicate. There are many general texts about the theory, principles and procedures of EIA (see, for instance, Munn, 1979; Clark *et al.*, 1984); and an excessive interest in methodologies and techniques has in our view tended to direct attention away from viewing the experience of EIA within the broad process of environmental planning. When reading the literature of EIA the impression is given that the philosophy and concepts are largely accepted wisdom – to find forceful critiques of EIA one must return to the literature of the mid-1970s (see for instance Brooks, 1976; Eversley, 1976). Emphasis is given to the description of case studies which illustrate procedural practice (Petts and Hills, 1982). The difficulty with all case studies is that experience is not simply transferable from one situation to another because the parameters used in the identification and execution of EIAs have been different in each case. Moreover, there is no consensus on how to define environmental impact assessment and the range of studies included under the umbrella of 'EIA' varies widely. A more promising line of inquiry than case studies is the institutional context within which specific development projects arise and EIA is made operational (for instance, O'Riordan, 1982). Yet the particular political and decision-making factors which influence why and how EIA has been used in different contexts have not been analysed; nor is much interest shown in the outcome of EIA upon planning decisions.

It is also relevant to consider the wider social and political implications posed by the adoption of EIA in the United Kingdom. In saying this we are aware that there is no consensus on what these are, nor on how any kind of judgements should be made. The standpoint from which we choose to evaluate EIA will depend on individual values about such broad considerations as the economy, society and the importance of environment to both. Should we evaluate the success of EIA solely in terms of the protection it affords the natural environment, as some ecologists might argue? Or is success measured by those developments which lead to a more equitable balance of social costs, as some social scientists might argue? Or is EIA rather a means of ensuring that externalities are more fully taken into account in investment appraisal decisions in both private and public sectors of the economy, as economists might argue? The interest of political scientists lies in whether EIA will bring greater openness of government. Shall we see an extension of public participation or merely a form of token consultation when EIAs are prepared? Will the institutions of government, particularly central and local government, operate more effectively once EIA is formally installed? Can we expect fewer delays in decisions on major public investment projects such as Stansted or Sizewell or will EIA (assuming no change in the public inquiry procedure) simply add to the lead time involved? In sum, we felt it important for contributors to address a range of issues:

(*a*) the economic, social, political and institutional context within which government and the private sector have introduced (or failed to introduce) EIA;

(*b*) the problems of legislating for environmental safeguards at a time of uncertainty in economic and political change;

(*c*) the changing nature of the public planning process and the application of EIA in forward planning and development control;

(*d*) approaches to EIA, especially: why a decision was made to adopt it and the utility of the results in the light of subsequent experience;

(*e*) the *post-hoc* evaluation of the lessons learnt from the decision to adopt and use EIA, with particular emphasis on the criteria being used to evaluate EIA.

Many books already emphasize the specialized methods needed to undertake EIAs. We direct this book as much to newcomers to the literature of EIA as to those politicians, businessmen, town planners and environmentalists involved in EIA practice. Although each chapter stands on its own, the structure of the book is progressive. It moves from a description and analysis of previous experience to future legislative changes and their implications for practice and methodology and concludes with an evaluation of the implications of EIA in the context of the more general political and institutional changes currently affecting the UK planning system.

The first three chapters are taken up with *post-hoc* evaluation of the lessons to be learnt from the decision to adopt and use EIA in rather different geographical and political circumstances. The role of EIA in the planning process is approached through a case study examination of the practical experience of planners working in local government, and is contrasted to the lessons we can learn when developers have used EIA. In the next two chapters Williams and Wood outline the main requirements of the EC legislation and how these may relate in practice to the UK system of development control. The opportunity for policy assessment may seem anathema to Civil Service procedure but government will need to be more accountable than it presently is for its own politically sensitive policies, especially in the spheres of energy and transportation policy. EC requirements will demand more sensitive government approaches to the formulation and evaluation of national and regional policy and new kinds of methodology for exploring the complex relationship between policies and projects, as Wathern and colleagues explain in Chapter 6.

The wider implications of EIA are treated by Clark, Herington, McDonic, Cope and Hills in the last four chapters of the book. Clark shows how the chance to use EIA allows us to examine our ideological commitment to planning in general and to appraise the implications for the UK planning system of introducing a more formal method of environmental assessment, including the possibility of widening the scope of planning powers. Herington argues that developing new and improved methods of assessment will do little to improve the quality of planning decision making if political decisions still take no account of environmental values. We need to be explicit about what planning actually is, and therefore how EIA can or might fit in, and why. McDonic, who at the time of writing is President of the Royal Town Planning Institute, foresees that the advent of the EIA Directive in 1988 will help strengthen the present planning system. Finally, Cope and Hills take a philosophical view of the spectrum of ideas underpinning EIA. Although their chapter was written in the late 1970s it is included in the collection because the arguments have proved remarkably robust. A 'comprehensive assessment' of policy issues remains a valid goal for governments to strive toward; even the advocates of EIA will need to widen their horizons to cover the full range of impacts demanded by an increasingly articulate democracy.

REFERENCES

Anderson, P. (1985) Habitat and landscape conservation: current strategies in the National Parks. *Ecos*, 6, pp.18–24.

Bidwell, R. (1985) The gap between promise and performance in EIA: is it too great? London: Environmental Resources Limited. Mimeo.

Blowers, A. (1984) *Something in the Air: Corporate Power and the Environment*. Oxford: Pergamon.

Breakall, M. and Glasson, J. (eds.) (1981) *Environmental Impact Assessment: From Theory to Practice*. Oxford: Oxford Polytechnic Department of Town Planning.

Breheny, M. and Hall, P. (1985) The strange death of strategic planning and the victory of the do-nothing school. *Built Environment*, 10 (2), pp. 95–99.

Brooks, E. (1976) On putting the environment in its place: a critique of environmental impact assessment, in O'Riordan, T. and Hey, R. D. (eds.) *Environmental Impact Assessment*. Farnborough: Saxon House.

Clark, B.D., Gilad, A., Bisset, R. and Tomlinson, P. (eds.) (1984) *Perspectives on Environmental Impact Assessment. Proceeding of the annual training courses, Aberdeen 1980–1983*. Dordrecht: D.Reidel.

Clark, B.D. and Turnbull, R.G.H. (1984) Proposals for environmental impact procedures in the UK, in Roberts, R.D. and Roberts, T.M. (eds.) *Planning and Ecology*. London: Chapman and Hall, p. 136.

Commission of the European Communities (1984) *The FAST Report, Eurofutures, the Challenges of Innovation*. London: Butterworth.

Cope, D.R. (1984) Radioactive waste management and land-use planning, in Cope, D.R., Hills, P. and James, P. (eds.) *Energy Policy and Land-Use Planning*. Oxford: Pergamon.

Countryside Commission (1986) *Annual Report 1985–6*. Manchester: Countryside Commission.

Council for the Protection of Rural England (1981) *Planning – Friend or Foe?* London: CPRE.

Department of the Environment (1988) Environmental Assessment: Implementation of EC Directive. Consultation paper. London: DoE, mimeo.

Dean, F.E. and Graham, G. (1977) Environmental Impact Analysis in the British Gas Industry. Economic Commission for Europe (ECE) symposium on the Gas Industry and Environment, Minsk, 20–27 June.

Ekins, P. (1986) *The Living Economy: A New Economics in the Making*. London: Routledge and Kegan Paul.

Eversley, D.E.C. (1976) Some social and economic implications of environmental impact assessment, in O'Riordan, T. and Hey, R.D. (eds.) *Environmental Impact Assessment*. Farnborough; Saxon House.

Fairclough, A.J. (1984) Environmental protection and industry – the European Community's approach. *Industry and Environment*, 7 (3), pp. 47–49.

Goode, D.A. (1984) Conservation and value judgements, in Roberts, R.D. and Roberts, T.M. (eds.) *Planning and Ecology*. London: Chapman and Hall.

Greater London Council (1985) *The Erosion of the Planning System*. London: GLC.

Gregory, R. (1971) *The Price of Amenity*. London: Macmillan.

Grove-White, R. (1984) The role of EIA in development control and policy-making, in Roberts R.D. and Roberts, T.M. (eds.) *Planning and Ecology*. London: Chapman and Hall, p. 148.

Hardin, G. (1968) The tragedy of the commons, in Blunden, J. *et al.* (1978) *Fundamentals of Human Geography: A Reader*. London: Harper and Row, pp. 76–83.

Heath, M. (1985) Deep digging for nuclear waste disposal. *New Scientist*, 1480, pp. 30–32.

Herington, J. (ed.) (1979) *The Role of Environmental Impact Assessment in the Planning Process*. Loughborough: Institute of British Geographers and Loughborough University Department of Geography.

House of Commons Environment Committee (1986) *Planning Appeals, Call-ins and Major Public Inquiries*. London: HMSO.

MacEwen. A. and MacEwen, M. (1982) *National Parks: Conservation or Cosmetics?* London: Allen and Unwin.

Mills, S. (1986) Roads run over conservation. *New Scientist*, 1496, pp. 44–46.

Munn, R.E. (ed.) (1979) *Environmental Impact Assessment: Principles and Procedures*. Chichester: Wiley.

Nature Conservancy Council (1986) *Nature Conservation and Afforestation in Britain*. Peterborough: Nature Conservancy Council.

Nuffield Foundation (1986) *Town and Country Planning – A Report to the Nuffield Foundation*. London: Nuffield.

O'Riordan, T. and Turner, R.K. (eds.) (1983) *An Annotated Reader in Environmental Planning and Management*. Oxford: Pergamon, p. 10.

O'Riordan, T. (1982) Institutions affecting environmental policy, in Flowerdew, R. (ed.) *Institutions and Geographical Patterns*. London: Croom Helm, pp. 13–14.

Pepper, D. (1984) *The Roots of Modern Environmentalism*. Beckenham: Croom Helm.

Petts, J. and Hills, P. (eds.) (1982) *Environmental Assessment in the UK: A Preliminary Guide*. Nottingham: University of Nottingham Institute of Planning Studies.

Porritt, J. (1984) *Seeing Green: The Politics of Ecology Explained*. Oxford: Blackwell.

Rees, J. (1985) *Natural Resources: Allocation, Economics and Policy*. London: Methuen.

Roberts, R.D. and Roberts, T.M. (eds.) (1984) *Planning and Ecology*. London: Chapman and Hall, p.100.

Rose, C. (1984) Acid rain falls on British woodlands. *New Scientist*, 1482, pp. 52–57.

Royal Town Planning Institute (1986) *The Challenge of Change*. London: RTPI.

Ryder, A., Bisset, R. and Kayes, P. (1986) Outline of the problems over assessments. *Planning*, 690, pp. 10–11.

Shoard, M. (1980) *The Theft of the Countryside*. London: Temple Smith.

Short, J. (1984) New pressures on London's Green Belt. *Geographical Magazine*, 56, pp. 90–92.

Slocumbe, D.S. (1984) Environmentalism: a modern synthesis. *Environmentalist*, 4 (4) pp. 281–5.

Stroud, D. (1985) The case of Duich Moss. *Ecos*, 6, pp. 46–48.

Strong, M. (1984) Major issues facing the conservation movement in the coming decade and beyond. *Environmentalist*, 4 (3) pp. 165–75.

Town and Country Planning Association (1985) *Sizewell Report: A New Approach for Major Public Inquiries*. London: TCPA.

Chapter One

Environmental Impact Assessment in the United Kingdom

R.G.H. TURNBULL

There was a clear post-war consensus in the United Kingdom about the need to protect rural and coastal environments from uncontrolled urban and industrial development. The well known wartime reports of Barlow, Scott and Uthwatt paved the way for the subsequent Town and Country Planning Acts, and the 1948 National Parks and Access to the Countryside Act included conservation measures and a system of state nature reserves. Since that time, the encroachment of towns and industrial development on the rural environment has continued and there is a growing demand for outdoor recreational and other facilities. Economic growth and urban development, reflected by the impact of offshore and onshore oil and gas discoveries in Scotland, has reinforced the case for protective and careful planning of the natural and man-made environment.

Many of the detailed aspects of protecting the environment from airborne and other forms of pollution are undertaken at local government level with the advice and guidance of the Industrial Pollution Inspectorate, River Purification Boards, the Health and Safety Executive and others. Overall planning and management strategy is needed to coordinate interests in the protection and the enhancement of the environment. Consultation between the many organizations and agencies concerned with planning the environment is now well established as part of the decision-making process. Well defined procedures are essential to ensure that economic development and environmental protection can and do take place within a reasonable timescale and in as mutually compatible manner as possible. In this context, environmental impact assessment has an important role to play in formulating and evaluating environmentally sound policies and plans, as well as a basic tool for the assessment of individual development proposals.

Environmental Policy

Although doubts over the formal adoption of environmental impact procedures continue to be raised, the earlier concern about unproductive procedural delays and costs, at a time when government was seeking to expedite and simplify planning procedures, has been unfounded.

The administrative implications of introducing a formal impact assessment system were examined first by consultants to the Department of Environment (Catlow and Thirlwall, 1977). Their recommended approaches were subsequently adopted by both public and private sector, and the government indicated that they wanted to see environmental impact assessment extended to major developments affecting environmentally sensitive areas. Planning authorities would be required to agree, at as early a stage as possible, the need for, form of, and methods of preparing an impact assessment, including the division of responsibility for carrying out the work. All interested parties were to be informed about the scope and nature of the analysis undertaken.

Inevitably, questions were raised about how the decision-making procedures would operate and whether or not planning authorities should extend the scope of assessment to cover issues specifically raised by interested parties. The adoption of the EC Directive on environmental assessment (Commission of the European Economic Communities, 1985) imposes new mandatory requirements in the use, content and conduct of assessment. The Directive firmly places the responsibility for preparing the environmental impact assessment and its content with the developer, contrary to current practice which advocates that the planning authority should do this with the assistance of the developer. The advantages of joint assessment are the reduced likelihood of duplicated effort and the avoidance of partial and biased assessment following from the developer being judge and jury of his own project. Moreover, some developers may be reluctant to provide project information or accept the importance of environmental matters thus risking conflict with planning authorities.

The Early Stages – Onshore Oil and Gas Developments in Scotland

Environmental impact assessments were largely the outcome of the National Environmental Policy Act (NEPA) in the United States which came into effect in January 1970 (Council on Environmental Quality, 1978). This Act required the production of environmental impact statements (EISs) for major Federal projects, and it stimulated research into ways of identifying and measuring environmental impacts. The legislation came into being more or less simultaneously with the first onshore effects of the offshore oil and gas activities in Scotland.

From 1970 onwards, a series of major and unfamiliar development proposals posed difficult problems for planning authorities and gave rise to specially thorough and positive methods of assessment within the development control system. The impact of these new energy proposals called for macro-scale planning; for short-term and long-term planning; for economic, land-use and social planning; for reactive planning; and for the creation of an overall framework within which others could develop and plan. It required a reassessment of planning procedures by central and local government.

The Scottish Office was first to provide guidance on how to approach the appraisal of the impact of oil and gas-related developments. It advised local authorities to ask questions about the proposal and to discuss methods of carrying out an impact analysis. The development control system and staff complements were strengthened; a variety of planning, environmental and engineering studies were commissioned by central government in conjunction with local planning authorities.

Between 1976 and 1980 research and planning activity was centred on the environmental implications of constructing a gas-gathering pipeline system in the North Sea with a landfall at the St Fergus gas terminal north of Aberdeen. The onshore proposal to extract natural gas liquids at St Fergus and to transport them to Nigg Bay on the Cromarty Firth, involved detailed engineering and environmental assessment of overland and undersea pipeline routeings; choice of alternative terminal and processing sites on the Moray Firth coast; and major land reclamation possibilities. The concurrent discovery in the Moray Firth of commercial quantities of oil some 25 km offshore in the Beatrice Field, and the intention of the developer to bring the oil ashore by tankers, led to considerable controversy about the technical and environmental issues involved. For the first time, the Department of Energy requested the offshore developer to commission an EIA to help appraise the technical feasibility and cost implications of an undersea pipeline to shore, and test the integrity of the proposed offshore tanker system. The study examined the potential effect of three modes of bringing the oil ashore on the marine and shore environment of some 300 km of coastline. This brought together a considerable amount of information and knowledge of the coastal conditions in the Moray Firth area.

By 1980, some thirteen major environmental assessments were carried out for oil-related development (see figure 1). Other projects included a hydro pump-storage proposal, alternative water catchment schemes, major road alignments and overhead electricity transmission routes. These studies were carried out in various ways. The first consultant's reports involved a regional assessment of alternative sites for concrete platform construction. Others were undertaken by the planning authorities concerned, sometimes employing consultants to cover specialized aspects such as noise, pollution, risk and hazard.

At the request of central and local government, developers commissioned reports from consultants. Some appraisals were aimed at anticipating possible developments rather than responding to specific projects. For the first time it was appreciated that EIA could become a tool for use in forward contingency

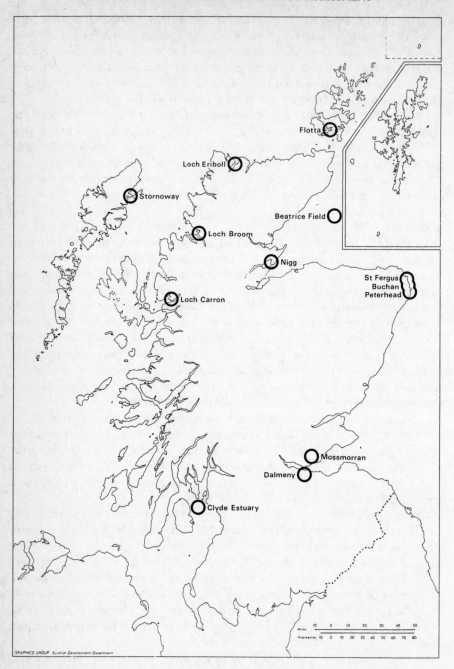

Figure 1. Location of impact analyses for North Sea oil developments.

planning through the selection of suitable sites for large-scale industrial development.

Developing Research Interest in EIA

The first comprehensive environmental survey and assessment on a sub-regional scale examined the likely impact of potential petrochemical developments (Economic Intelligence Unit, 1975). The study was commissioned in 1972 and was extended in 1978 to include an engineering and environmental feasibility study of potential harbour sites (Rendall, Palmer & Tritton, 1978).

Other important research followed the commissioning by central government of the *Manual for the Assessment of Major Development Proposals* (PADC, 1981). This included the identification of pipeline landfalls and associated gas terminal sites down the east coast of Scotland; the onshore effect of a proposal for a gas-gathering pipeline in the North Sea. The experience gained from the various studies led the Scottish Office to publish a series of discussion papers on the subject of pipeline landfalls, oil terminals, onshore oil and gas pipelines, landscape assessment, environmental monitoring by remote sensing, and a series of planning information notes on different types of petrochemical plants and processess and their environmental effects (Scottish Development Department, 1974*a*, 1974*b* and 1977).

Environmental impact assessments concentrate on identifying main issues and avoiding over-elaboration in the use of techniques and methods. There has also been a recognition that environmental considerations should be integrated into the various levels of planning and decision-making framework. There are, however, continuing differences of view as to the form such integration should take.

The prospect of using EIA in the statutory planning system came a stage closer after the Standing Commission on Energy and the Environment (set up to advise government on the interaction between energy policy and the environment) concluded, 'that a mandatory system of formal EIA would risk undue delays and costs in the planning procedure' (Commission on Energy and the Environment, 1981). The preference was for a wider deployment of EIA within the existing flexible approach to project assessment.

Administrative Responses

In 1975 the Dobry Report on the Development Control System advocated that in the case of specially significant development proposals the planning authority should be able to insist that impact studies be submitted by the developer. A year later, the Fifth Report of the Royal Commission on Environmental Pollution (1976) made a number of recommendations for linking pollution control mechanisms into the planning process. It endorsed the Dobry proposal that developers should provide an assessment of the effects of air, water, waste and noise pollution of certain major projects.

The House of Lords Select Committee on the European Communities (1981) welcomed the proposal for an EC Directive which would require environmental impact assessments to be carried out before planning consent was granted for special classes of development project. This view was reached after detailed consideration of written and oral evidence being taken from central and local government, industry, consultants, national agencies and associations. The Committee rejected the view that there should be a Code of Practice rather than a Directive on the ground that a Directive is appropriate because its implementation would be obligatory on all Member States.

Arguments for and against the Directive were considered by the EC Economic and Social Committee, the Environment Committee and the Legal Affairs Committee. The first unanimously approved the proposal that the Directive should lay down common principles for EIA, and that a co-ordinated evaluation procedure would bring certain advantages, not only for planning authorities which would have a better understanding of projects, but also for the developer through the simplification of administrative procedures. Attention was also drawn to the problems raised in placing responsibility on the proponent of a project to prepare EIA documentation for the appropriate planning authority. The problems involved were confidentiality of information, the cost of preparing an EIA, the time needed to obtain planning permission, and the submission of alternatives to the initial plan. The Environment Committee also supported the Directive proposals and expressed the hope that greater openness in the planning and approval of projects could lead to better relations between the public, developers and the authorities. Time consuming and costly appeal procedures, in their view, could be avoided by participation of the public in the EIA process at an early stage.

National Planning and Environmental Assessment

The White Paper on 'Land Resource Use in Scotland' (House of Commons, 1972) for the first time, provided a remit for the Scottish Office to undertake the formal activity of National Planning as distinct from the rest of the United Kingdom. This occurred more or less simultaneously with the introduction of the new Development Planning and Development Control system in 1972, and the subsequent reorganization of local government in Scotland in 1975. The first step in the protection of the environment was the issue of *Coastal Planning Guidelines* (Scottish Development Department, 1974b) which divided the coast of Scotland into preferred development and preferred conservation zones. It provided national guidance to both planning authorities and potential developers as to the relative difficulties of obtaining planning permission on the coast.

Prior to the publication of the guidelines, the Scottish Office in collaboration with the Countryside Commission and the Nature Conservancy Council commissioned consultants to carry out a detailed survey and evaluation of the environmental features of the coastline. The findings were published in 1975, as a volume of maps and descriptive information titled, *The Coast of Scotland: Some Recently Collected Survey Material.*

A further series of national guidelines and land-use summary sheets covering petrochemical developments, large-scale industrial sites, conservation areas, agricultural land, aggregates, landscape and recreation were issued (Scottish Development Department, 1977). Planning authorities were encouraged to 'have in mind', the broad national policies indicated in the guidelines when preparing their development plans and in dealing with environmental issues. Reference was made to the earlier coastal planning guidelines which would continue to form a basis for policy. The guidelines contained suggestions for more positive action by planning authorities, and site selection criteria for determining the potential for petrochemical development in their respective areas, in advance of specific proposals. A number of planning authorities responded by commissioning special studies including preliminary environmental assessments. The results indicated that it was possible, at a very early stage, to identify the main environmental impediments of pre-selected sites for petrochemical development (Grampian Regional Council, Banff and Buchan District Council, 1980). This has resulted in a much better understanding of the environmental issues and operational factors involved in the advance selection of industrial sites. The main elements in the selection process were safety, pollution, land acquisition, visual amenity, utilities and communications. The availability of guidelines covering agricultural land, forestry, landscape and recreational interests were instrumental in limiting the areas of search and provided safeguards for the protection of specific areas of land with special characteristics.

The successful use of environmental impact assessment in the pre-selection and allocation of appropriate land uses and in plan-making is largely seen as being dependent upon the existence and practice of well-established planning and pollution control systems. Where such systems are non-existent, the problems of using EIA are largely academic. The support given by national planning guidelines lies in their value in providing broad base-line information on the environment and in identifying broad policy.

The continuing debate at national and international levels about the nature, magnitude and cause-effect relationships of various pollutants on the environment has focused interest on the ways and means pollution problems are tackled at different levels. The main outcome of these deliberations has been the acceptance of international agreements, conventions, directives, and other legislation. While many of the pollution problems are universal, there is a general acceptance that each nation has its own special problems and requirements dictated by climate, geography and economic circumstances. The focus of concern has been the long-range transport of pollution between countries which led to the provision in the EC Directive on Environmental Impact

Assessment for consultation procedures in dealing with transfrontier pollution issues. In essence the developer and competent authority in the Member State initiating the project have responsibility for deciding the nature and extent of the information to be provided to the 'affected' Member State for comment and for deciding whether or not another Member State may be affected. However, submitting descriptive information and asking for comment only has limited value since no Member State who may be exposed to transfrontier pollution has a right of veto, nor is given the opportunity to discuss and contribute to the preparation of the EIA.

EIA and Local Planning Control Issues

A wide variety of development types of varying complexity and size are undertaken by both public and private developers. These projects are subject to a range of controls under which permission is needed from the responsible public authority before they can be undertaken. Which permissions are necessary differ with the scale and nature of the project, layout and design, process, location and other factors. Not every authorization procedure is concerned with the environmental effects of the proposed development. In the energy industry certain types of development used for the production and distribution of electricity and gas require authorization by the Department of Energy. The authorization carries with it deemed planning consent.

Control procedures are comprehensive and most decisions are taken at local authority level against the policies and proposals contained in the approved structure plan and local plan for the particular area. Both structure and local plans are the result of extensive public consultation and detailed consideration of comments, ideas and opinions expressed by a wide range of interests. They are not simply a county/regional or district council document but rather a planning framework for agencies in both the public and private sectors. The agencies include: the Nature Conservancy Council, the Countryside Commission, the Royal Society for the Protection of Birds, the National Trust, the National Boards and others.

Before a decision is taken to authorize or approve particular projects, there is further provision for public consultation and participation in the decision-making process. In addition, there is an abundance of planning and other legislation, operated by a number of special environmental protection organizations to ensure that after construction is completed, the environment will continue to be safeguarded and the public protected.

In the decision-making process the role and content of environmental impact assessments have varied to meet the needs and circumstances of particular projects and locations. Over the years, they have become more sophisticated in their presentation and content and, in the case of 'hazardous industries', they have been extended to include risk and hazard evaluations and operational studies prior to the commissioning of hazardous process. This has posed a number of questions about the status and value of environmental impact assessments in relation to the granting of provisional, outline and full planning permissions.

The purpose of applying for provisional or outline planning permission is to enable an applicant to establish whether, on a given site, a particular type of development will be permitted, without at the initial stage having to prepare and submit detailed plans of his proposal. The idea is that the planning authority agree to the development proposal in principle and such development may take place subject to the submission of satisfactory details. Such details, known as 'reserved matters' may relate to siting, design or external appearance of any building to which the outline permission relates, or means of access to it, or the landscaping of the site which it occupies. In dealing with reserved matters the planning authority may not revert to questions of principle in order to refuse the full application.

When outline planning applications are accompanied by well presented EIAs together with detailed site plans, it is often unclear whether the details are merely illustrative information or part of the application. In such circumstances, the local planning authority granting the outline planning permission, with or without reservations, must accept the possibility of major subsequent amendments to the proposed development. When this occurs, a judgement will be required as to whether or not the application should be resubmitted with a new or amended environmental impact assessment. It must also be recognized that issues concerning pollution, waste disposal, risk and hazard to surrounding communities are inappropriate topics for any reservation in respect of an outline consent. Other issues affecting the use of EIA, are the availability of information, problems of confidentiality, consideration of alternative sites, and how to deal with plant processes at a very earlier stage of the project design.

From time to time, the legal status of an EIA accompanying a full and/or an outline planning application has been discussed without any clear decision being taken. Under the present control system the tendency has been to regard environmental impact assessments as supporting information to help the decision-making process. However, if it is to become a statutory part of the project application under the provisions of the EC Directive, it will place a commitment on the developer to implement the stated proposals and policies and related environmental provisions.

The Evolving Use of Environmental Impact Assessments

Within the framework of the present UK planning and control system, the responsibility and cost for undertaking assessments is usually decided between the planning authority and the developer concerned. In a number of cases, assessments have been undertaken by both parties to meet their respective needs and interests. In other cases, local authorities have negotiated a financial contribution from the developer towards the preparation of an 'independent' environmental impact assessment. This is of some importance to smaller authorities with limited staff and financial resources.

The Manual for the Assessment of Major Development Proposals was designed for flexible use by both the local planning authority and developer. Its use does not involve any alteration in statute, and the structured method of appraisal for major projects is largely based on planning authority experience gained in the application of environmental impact assessments. In addition to being used on single projects it has been readily adapted to other tasks in forward planning.

From a variety of studies undertaken so far, two main sets of conclusions may be drawn. First, EIA is here to stay and can be used in a wide range of situations and circumstances. The scope, organization and emphasis of each assessment can vary with the project, whether it is a remote possibility or a highly probable development, and whether one or more developers are involved. This also applies to linked linear projects such as oil and gas pipelines and overhead transmission lines which extend over large geographical areas. The common purpose in each case is to assess the effect of a proposed project on the physical, social and economic environment of the host area. The basic justification being that this approach leads to 'better' decisions than would otherwise occur.

Second, the critical factors in producing a satisfactory assessment within a reasonable timescale require a systematic approach, adequate information about the project and environmental factors, and expertise to identify and predict potential impacts. To reach sound judgements there must also be an awareness of the external context of national, regional and local attitudes, and of the appropriate administrative procedures.

The adoption of a systematic approach may not reduce the volume of work involved in an environmental impact assessment but it ought to ensure a more efficient use of time and resources and avoid abortive effort. This is of particular importance in the collection of information about the project, the host area and in the selection of impacts. Problems will inevitably arise in identifying and measuring the significance of direct and indirect impacts, singly and in combination. For example, the indirect and long-term environmental health impacts may be many years in becoming apparent. Again, subtle changes in values and attitudes of different groups of people towards particular types of developments may prove impossible to predict or describe. These and other difficulties in carrying out environmental assessments must be recognized and made known to the decision-makers and the public at large.

In retrospect, the development control system has proved sufficiently robust to deal with many unfamiliar development projects, and flexible enough to accommodate environmental impact assessments into the decision-making process. The procedures followed have been broadly in line with the principles set out in the EC Directive, but they have fallen short of the detailed requirements for public consultations and in the coverage and contents of an assessment. The ultimate test of the adequacy or otherwise of an environmental impact assessment can only be applied after the particular project becomes operational and the impacts have been monitored and assessed at agreed intervals of time.

European Directive on EIA

The European Community Environment Action Programme established in 1973, and extended in 1977 and 1982, ran to 1986. The action programmes covered many aspects of environment policy including the basis for the Directive on the assessment of the environmental effects of certain public and private projects (Commission of the European Communities, 1985). The Directive introduced common principles applicable to the main obligations of developers; the types of project to be subject to assessment procedures; the content of the assessment; and the environmental factors to be taken into account. It draws attention to the decision-making and planning process with respect to projects, land-use plans, regional development and economic programmes. Priority, however, is given to the introduction of the principles of assessment to projects. Member States are required to take measures to comply with the Directive within three years of its notification in 1985.

The fourteen Articles and three Annexes of the Directive provide for a mandatory assessment of certain major development projects and an assessment of others at the discretion of the Member States. The selection of the latter being on the basis of criteria and/or thresholds to be established by the Member States, who may also determine whether or not the project should be subject to a full or simplified assessment or be exempted. No guidance is given on the establishment of criteria and thresholds or as to the content or coverage of a simplified assessment. A major obligation is placed on the developer to provide the necessary information and assessment for decision by the planning authority.

To meet these requirements the Directive raises a number of technical and administrative issues affecting the present flexible approach to EIA in the United Kingdom. These include the need to clarify the difficulties involved in the formulation of meaningful criteria and thresholds for the selection of projects for assessment; the distinction between full and simplified assessments; and the procedural and technical difficulties of dealing with transfrontier pollution between Member States. Again, the responsibility placed on the competent authorities to provide guidance and assistance to the developer in the preparation of an assessment may, or may not, be in conflict with their own interests. Little or no guidance is given as to the review and decision-making mechanisms to be followed in dealing with private and public projects.

The preselection of criteria and thresholds to determine whether or not a development is likely to give rise to significant environmental impacts in all locational circumstances will be difficult, if not impossible. It is also difficult to see how criteria and thresholds can be adjudicated upon at EC level to satisfy all

Member States. The significance of environmental impacts are largely local phenomena which are dependent upon existing levels of pollution and other conditions, consequently a flexible approach to the formulation of criteria and thresholds is required. For example, central government could set out the broad principles for establishing criteria and thresholds which would be interpreted by the competent authorities in light of their local circumstances.

Differences between current UK planning control and other authorization procedures and those of the Directive are likely to require legislative and administrative changes. These might include some legislation to bring, for example, agriculture and other uses, which do not require planning consent at present, under planning control. Other cases, such as development by statutory undertakers, would require changes to the use classes and permitted development under the General Development Order (GDO). The basic requirement being the need to ensure that any enacted legislation, amendments to the GDO, and implementation by Statutory Instruments or Circulars would result in compliance with the provisions of the Directive. The responsibility for implementation rests with the Department of Environment who published a consultation document in 1988 setting out proposed procedures.

View of the Future

There is no universal 'cook-book' recipe for assessing the environmental impacts of development projects in all environmental settings. The present state-of-art ranges from improvements or new initiatives in the relatively precise and objective measures of air and water quality to more subjective measures of landscape and visual quality. Research and development work continue on study technology and the adaptation of the many available techniques and methods to different situations and circumstances.

Over the past decade or more, world-wide interest in the protection of the environment has resulted in the production of many thousands of EIAs of varying length and quality. Little work has been done, however, to test past findings and recommendations through the application of standard scientific procedures to assess present performance and thereby to effect future improvements. Several components are required: better procedures for impact monitoring, auditing the consequences of operational development, and appraising the effectiveness of measures to prevent harmful environment impacts.

The history of environmental impact assessment development in Britain emphasizes the value of learning by experience. Research into project appraisal methods for a single development was extended into forward contingency planning for large-scale industrial and petrochemical developments. National planning guidelines made an important contribution to environmental planning and development control in Scotland. The timely implementation of the EC Directive, if it were coupled with a programme of post-audits of previous EIAs, would help identify gaps in the earlier approaches. It should also provide know-ledge and understanding which could be applied to modify or supplement future environmental action programmes, at least to the extent of recognizing trends, possible environmental hazards, and 'probable or most likely' environmental impacts.

REFERENCES

Catlow, J. and Thirlwall, C.G. (1977) *Environmental Impact Assessment*. Department of Environment, Research Report, No. 11. London: HMSO.

Commission on Energy and the Environment (1981) *Coal and the Environment*. London: HMSO.

Commission of the European Communities (1985) *Council Directive Concerning the Assessment of the Environmental Effects of Certain Public and Private Projects*. EEC Brussels.

Council on Environmental Quality (1978) *Regulations for Implementing the Procedural provisions of the National Environmental Policy Act (NEPA)*. Washington DC.

Dobry, G. (1975) *Review of the Development Control System*. London: HMSO.

The Economic Intelligence Unit Ltd (1975) *Buchan Impact Study prepared for Aberdeen County Council and the Scottish Office*. London.

Environmental and Resources Consultancy Fairey Surveys Ltd (1978) *Environmental Monitoring by Remote Sensing*.

Grampian Regional Council & Banff & Buchan District Council (1980) *Contingency Plan for Petrochemical Industries*.

House of Commons Select Committee on Scottish Affairs (1972) *The White Paper on Land Resource Use in Scotland*. (House of Commons Paper 5/11 session 1971/2) London: HMSO.

House of Lords Select Committee on the European Communities (1969). Environmental Assessment Projects (11th Report 1980–81, House of Lords Report 69) London: HMSO.

PADC (1981) *A Manual for the Assessment of Major Development Proposals*. London & Edinburgh: HMSO.

Rendell Palmer & Tritton (1978) *Harbour Feasibility Study, Initial Assessment of Sites*. Prepared for Grampian Regional Council.

Royal Commission on Environmental Pollution (1976) Fifth Report, *Air Pollution Control: An Integrated Approach*. London: HMSO.

Scottish Development Department (1974*a*) *Appraisal of the Impact of Oil Related Development*. DP/TAN/16. Edinburgh. SDD.

Scottish Development Department (1974*b*) *Coastal Planning Guidelines*. Edinburgh: SDD.

Scottish Development Department (1977) *National Planning Guidelines for Petro-chemical Developments, Large-Industrial Sites, Agriculture, Conservation, Forestry*. Edinburgh: SDD.

Scottish Development Department (1978–80) *Planning Information Notes* (A, B, C & D Series). Edinburgh: SDD.

Chapter Two

Environmental Impact Assessment in Forward Planning: The Cheshire Experience

ALAN BELL

This chapter describes a case study carried out by Cheshire County Council into the development of a Local Plan for a major expansion of one of Britain's largest oil refineries and petrochemical plants. It outlines the approach adopted and explains the way in which environmental assessment was applied within a changing forward planning framework. The lessons learnt from using EIA within local government, and the advantages and disadvantages of this form of assessment are considered in the conclusion.

The study borrowed heavily from Scottish studies of the likely impact of new oil and petrochemical installations and the approaches suggested in Department of Environment research reports numbers 11 and 13 (Catlow and Thirlwall, 1977; Clark *et al.*, 1976). The scope, content, and approach adopted therefore owed a considerable debt to this original conceptual work on EIA in Britain.

Context

The Mersey Marshes Study was the culmination of several inter-related planning problems which developed along the south bank of the Mersey estuary over the last fifteen years (figure 1). The main problem arose from the need to accommodate the continued growth of the Stanlow oil refinery and petro-chemical complex. This major industrial complex originated in the late 1920s as an oil distribution depot, expanded to encompass bitumen production and oil refining, and gradually developed into an industrial complex including two oil refineries and a number of ancillary plants which have a land take of more than 809 hectares.

In the late 1970s planning authorities were faced with a large number of applications for relatively small developments. By 1979, following a burst of very

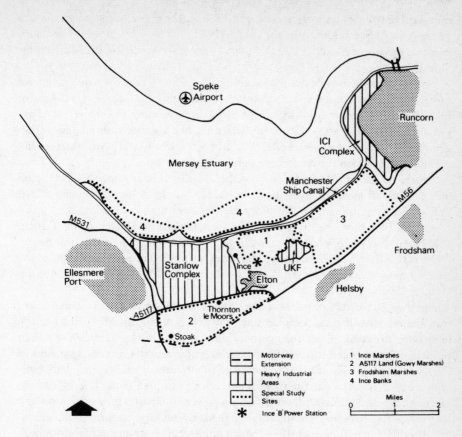

Figure 1. The Mersey Marshes Study Area.

rapid growth, development had reached the natural boundaries of the existing site which is formed by the Manchester Ship Canal, the villages of Ince and Elton, the A5117, and the town of Ellesmere Port (see figure 1). The planning authorities had to decide whether and where to release further land to accommodate continued development.

The broad flat banks of the Mersey estuary offered a number of sites capable of accommodating further heavy industrial development, all with excellent potential motorway and Manchester Ship Canal access. Site 1 and a substantial part of Site 2 were owned by existing firms for their future expansion needs, and Site 3 had been identified by the Department of Industry as one of the North West's major potential industrial sites. Site 4 had no comparable history of development interest largely because it was on the 'wrong' side of the Ship Canal and a large part of the potential site (in total about 1000 hectares) was subject to flooding at high spring tides. The area known as Ince banks is remote from existing settlements and thus appeared to be an ideal location for hazardous industry, despite its important ecological status (the entire Mersey estuary was made a Site of Special Scientific Interest in 1976), the likely high cost of development due to the need for reclamation, and the costs of bridging the Manchester Ship Canal.

Several other factors had to be taken into account. One issue was the impact of any industrial complex on nearby residential populations. The villages of Ince, Elton, Thornton-le-Moors, and Stoak were within or adjacent to these potential expansion areas. Elton, in particular, had seen substantial population growth (in 1979 there were still 16 to 20 hectacres of land with outstanding planning permission for housing development).

Another issue was the economic context. National policy encouraged the oil and gas industries for their use as chemical feedstocks, thus maximizing added value and employment in Britain and, incidentally, expanding the heavy chemicals sector of the economy (National Economic Development Office, 1977). As the North West has a very large share of Britain's existing heavy chemicals activity, national policy could not be ignored in any assessment of the case for new sites in the North West, the Mersey Valley in particular (Cheshire County Council, 1982).

Another factor has been the local experience of development control and the limited effectiveness of planning powers as a means of protecting the environment, even in relation to individual applications. An example which immediately preceded the Mersey Marshes study was the recent expansion of Shellstar at Ince. The Shellstar fertilizer plant (now U.K.F. Ltd.) had been allowed as a major departure from the development plan in the late 1960s but, although it was a modern plant, it had a record of unsatisfactory operation and had given rise to many complaints particularly relating to noise and atmospheric pollution. Consequently, when an application for a major extension was submitted, the County went to great pains to try to ensure that the environmental quality of the area was protected. The application was 'called in' and at the inquiry in January 1976 the County Council suggested a list of conditions relating to noise, pollution control, monitoring, etc, all but one of which had been agreed with the firm before the inquiry.

Planning permission was subsequently granted (though the decision was not received until May 1977) but even though the Minister's technical assessor had recommended that *all* the County's conditions should be attached, the Minister ignored the recommendations and imposed his own set of conditions which made no reference to control or monitoring, feeling that this should be left to the Alkali and Health and Safety Inspectorates, and the District Environmental Health Officers.

There was a further complication. The area is split administratively between three different District Councils with the existing complex and Sites 1 and 4 in one district while Sites 2 and 3 are in others. Local Plans are usually district matters, but in this case a County-prepared Local Plan was needed to overcome the difficulties presented by these administrative boundaries.

The policy conflict inherent in these expansion and development issues came to a head during the preparation of the County Structure Plan when the fundamental conflicts between economic development and the protection of environment and amenity posed serious policy-making problems. The strength of local public opinion showed quite clearly the importance attached to the question of heavy chemical expansion, and the draft Structure Plan, which was submitted to the DoE in spring 1978, dealt with the Mersey Marshes question in the following manner:

Site 3 was included in the modified North Cheshire Green Belt following the decision that no development should take place there, and air quality guidelines were suggested for the County which would provide the planning authority with a yardstick against which development control decisions could be taken and also give a means of controlling any cumulative build up of pollution. Site 1 was allocated for further industrial development (subject to suitable measures being taken to safeguard the amenity of local residents) so that the existing firms would have some expansion land available, but the future of Sites 2 and 4 was committed to more detailed examination in a special study.

The proposal for a special study was endorsed at the Examination in Public on the Structure Plan, though as a Local Plan rather than a Structure Plan amendment, while the proposal to put Site 3 into the Green Belt, and the idea of air quality guidelines attracted considerable opposition from industry and other statutory bodies and was not accepted by the Secretary of State.

Working Methods

The Mersey Marshes Study was run by Cheshire County Council, the work being carried out by a small full-time team of three professional officers who could call on help and advice from a number of specialists. Work was 'sub-contracted' to other specialists within the County Council including archaeologists, ecologists, architect planners, environmental health staff, valuers, highways and transportation staff, soil engineers; a small consultancy budget was used to buy specialist expertise not available within the County Council. The style of approach is illustrated by the following example.

The Mersey Marshes Study involved an extensive and elaborate system of consultation both at the stage of the original brief and subsequently when more specialist advice was required. Among the bodies involved were the Health and Safety and Alkali Inspectorates, the Manchester Ship Canal Company, the Mersey Docks and Harbour Board, the North West Water Authority and the Hydraulic Research Station. The public were also involved. The brief was discussed with local residents groups and environmentalists. The reports of survey were released for public debate in the winter of 1979 and later stages followed the statutory requirements for public participation in the Local Plan.

The management system which evolved was rather complex, partly because the scale and nature of the research did not fit easily within normal administrative systems. The Mersey Marshes Study was officially termed a County-prepared Local Plan. A special Local Plan steering group was set up to manage the study which included representatives of the four District Councils directly affected as well as the County Council. This was an officers group which reported to elected representatives at both District and County level through the normal system of committees. In addition, two special advisory groups were used to help with the development of the work: an Industrial Consultative Panel to comment on industrial policy, and a Members Panel which gave County Councillors the chance to comment in detail on reports, policies or proposals and ensured that elected members were fully informed.

Work on the study began towards the end of 1976 – it was intended to complete it within eighteen months to two years. However, the work was interrupted by preparations for the Examination in Public into the County Structure Plan (which involved the same staff) and the momentum did not pick up again until Spring 1978. By the following year, the study had finally reached its 'report of survey' stage and it was hoped that the draft plan would be completed by the summer of 1980.

The Report of Survey consisted of three main reports (see Appendix). They brought together as much factual information as possible in order to establish a baseline against which development proposals could be sensibly judged. Wherever possible the information was either quantified or directly descriptive, though inevitably findings on some topics depended upon value judgements. The reports provided a much more complete picture than would have been obtained if normal development control processes had been applied. Several interesting issues emerged.

(i) Pollution

Air pollution was the most thoroughly investigated of the environmental factors. It was already a live issue when the study began and was obviously one of the areas about which there was most local concern. The locals, mainly the action groups from the villages of Frodsham and Helsby which stand on a ridge above Frodsham Marshes, contended that topographical and climatic conditions combined to cause the build up of high concentrations of pollution in the lee of the ridge, and that chemical emissions, particularly of lead and ammonium nitrate were having a serious effect on health in the village.

A survey of air pollution was mounted to test these claims. It included a deposit gauge survey over a period of one year, with gauges placed at sites around the Stanlow complex to measure the deposition of fifteen specific pollutants including particulates, a number of organic compounds, and heavy metals. That survey was backed up in its final month by a parallel study of four sites involving a more sophisticated system of air sampling so that the concentration of pollutants (measured in parts per million) could be identified in relation to the volume of air. The substances measured in the second survey were sulphur dioxide, nitrogen oxides, ammonia, and ammonium and nitrate ions, these five being chosen in the light of the deposit gauge survey as the most likely to give rise to problems. A SYMAP programme was then developed to interpret the results from the twenty survey points and to produce maps showing the probable distribution of pollution across the area as a whole.

The results of the survey showed a concentration of sulphur dioxide around the refinery itself, though levels appeared to be decreasing over time and the effect was very localized. Other firms linked to the complex were seen to produce concentrations of specific pollutants, the most serious of which were the ammonium and nitrate emissions centred on the U.K.F. fertilizer plant. At the eastern end of Frodsham Marshes the ICI complex produced concentrations of some pollutants – mercury and fluoride being the most serious. However, despite the fact that levels were much higher than in the surrounding country-side, the concentrations fell well within national Threshold Limit Values (TVLs).

The question of lead in the environment was also of some concern as Associated Octel produced lead additives for petrol in an ancillary plant linked to the Stanlow complex. However the deposit gauge survey showed lead concentrations to be no higher than those found commonly in rural areas.

In addition to the monitoring of pollution levels, attempts were also made to identify the possible impact of existing emission levels on human health and ecology in the area. Comparison of available mortality and morbidity statistics for local authority areas in the vicinity of Stanlow with those for the nation as a whole showed no evidence of higher death rates for respiratory disease in the study area; moreover, they showed a national trend of gradual improvement which was reflected in the study area despite the intensification of chemical activity. Further, the routine examination of water supplies and milk from animals grazing on the Mersey Marshes did not show any problems of contamination which might have affected human health.

A survey carried out by Liverpool University's Environmental Rehabilitation Unit did demonstrate some damage to vegetation caused by chemical agents. Most of this could be attributed to pollution from traffic exhausts, with the exception of land immediately adjacent to the U.K.F. fertilizer factory, where there had been some damage to hawthorn hedges, and a barley field to the north of the plant which was quite severely affected. There were no wider ecological effects associated with the petrochemical plants, even though refinery and petrochemical activities have been located in the area since the 1920s.

This conclusion was borne out by other ecological evidence. For example, the size of the wildfowl populations of the Mersey Estuary increased dramatically in the late 1970s despite the very polluted air and water conditions. Today the estuary is second in importance in Britain as a wintering area for wading birds.

Although the general evidence suggested strongly that the petrochemical complex did not pose a major threat to human health, the possibility of more localized or longer term problems could not be ruled out. A further inquiry into the geography of mortality in the region was carried out to identify health problems which might be associated with particular pollutants. This work was essential to meet the public's expectations of the exercise.

(ii) Hazard

Cheshire County Council's interest in chemical hazards predated the start of work on the Mersey Marshes Study. The chemical consultants Cremer and Warner were commissioned after Flixborough to report on the processes, problems and potential hazards related to each of Cheshire's chemical plants. Their report, produced in 1976, together with the comments of the Health and Safety Executive, formed the basis of our assessment of hazard in the Mersey Marshes Study. Unfortunately, a more detailed hazard assessment was needed for the Local Plan work – in part because the Health and Safety Inspector did not always agree with the findings of Cremer and Warner.

The Cremer and Warner survey showed that the most serious potential hazards would arise within the complex, i.e. in parts of the Shell and Associated Octel plants. Large volumes of inflammable liquids and gases were stored on the site giving rise to risk of fire and explosion; a number of toxic chemicals were also handled, with the potential risk of the release of a vapour cloud which could settle at ground level and be carried along by the wind. The village of Thornton-le-Moors was located in a high hazard area. Elton, though to a lesser extent, was subject to a level of hazard risk which would make it prudent to refuse permission for any further building on its westerly side (figure 1).

Shell (who refused to co-operate with Cremer and Warner at the time the survey was conducted) disputed these findings. They were supported by the Health and Safety Executive who stated that no existing residential properties were subject to unacceptable risk and that Cremer and Warner's view that Thornton-le-Moors was an area of high hazard risk was unacceptable. They suggested that the need for renewal over the next few years would allow some of Shell's existing plant to be resited further away from Elton.

Another aspect of the research into hazard involved the question of buffer zones, or *cordons sanitaires*. Standards varying between ½ km and 2 km are in use in other countries, and have even been recommended in other parts of Britain, but the local office of the Health and Safety Executive felt that in the Stanlow area each application should be taken on its merits, and that buffer zones of more than 250 metres would not be justified.

The limited information on risk provided by existing operators made it unlikely that the next stage of work into the potential hazards of new sites would be successful.

(iii) Economic Impact

At the Shellstar Inquiry (and in several other planning applications) a number of claims and counterclaims were raised on the scale and nature of the economic benefits derived from the development of new large-scale chemical plants. Protestors offered two main arguments:

1. That good industrial land is wasted by capital intensive development because employment densities are so low (often yielding only one job per acre).

2. That petrochemical development produces a much greater pollution and hazard penalty than general industry.

However, others from industry claimed that large-scale petrochemical industry stimulated the regional economy, was a major wealth creator, and supplied raw materials to many industries, thus generating considerable spin-off employment.

Research showed the chemical industry was of fundamental importance to the regional economy, providing 47,000 jobs in heavy chemicals (more than 30 per cent of Britain's total employment in these sectors), nearly 150,000 in 'chemicals' as a whole, and over 250,000 jobs in total, if other closely linked sectors normally outside the Standard Industrial Classification (SIC) definition of 'chemicals' are included. This figure represented about 20 per cent of the North West's total manufacturing employment, and demonstrated a degree of dependence on the chemicals sector only rivalled by the Northern region.

The importance of chemical output to the national economy was also beyond dispute, the industry being one of Britain's top three money earners in 1977 with a balance of payments surplus of over £1400m. The chemicals sector (as defined in the 1968 SIC) accounted for 12 per cent of all exports and more than 25 per cent of Britain's trade balance in manufactured goods. The North West produced about a quarter of Britain's total chemical output; a contribution of about £350m to the national balance of payments in 1977.

In the North West the traditional chemical sectors (inorganic and miscellaneous heavy chemicals, of soaps and detergents, and dyestuffs and pigments) are still strongly represented and linked to the region's other traditional industries, textiles and glass and paper manufacture. Recent developments are strengthening the chemical sector, particularly 'organic' chemicals, oil-refining, heavy general chemicals, synthetic resins and plastics materials. However, these industries do not seem to be so closely integrated, at least in terms of location, with the downstream sectors which use their products as the traditional industries were. In fact, although the North West produced 22 per cent of Britain's plastics intermediates it had only about 11 per cent of national employment in plastics finishing.

The impact of new heavy chemical development on the local economy might be considerable. A major new complex could provide up to 2,000 permanent jobs with perhaps a further 1,500–2,000 temporary jobs during the construction phase. Even a smaller scale plant producing intermediate chemicals could employ 500–600. Other local benefits would include rates income (which can be substantial from capital intensive plants of this type), and a substantial local multiplier of two new indirect jobs for every new job on a chemicals or petro-chemicals plant.

At the regional scale the benefits of new development were unlikely to have much impact. Only a limited number of plants would be built and even on an optimistic assessment the new employment which they would generate was unlikely to exceed 5,000 jobs. Growth of this scale would probably be insufficient to offset the losses incurred by continued decline in the existing heavy sector (losses would be mainly in the 'inorganic' and 'miscellaneous' sectors) and in any case represented less than half per cent of the region's total employment. Economic spin-off was also unlikely to be significant at the regional scale. Although recent trends showed an increasing concentration of heavy sector activity in the North West this was unlikely to lead to expansion in the high growth downstream sectors such as pharmaceuticals and plastics finishing.

(iv) Possible Reclamation of Ince Banks

Ince Banks is the area of land identified as Site 4 on figure 1. It is an area of salt-marsh built up by the deposition of silt within the Estuary, occasionally flooded by very high tides, but nevertheless dry enough to support cattle and sheep for most of the year. The area was included as a possible alternative to the landward sites following public consultations on the County Structure Plan. Arguably, Ince Banks would be a superior site from the point of view of hazard and nuisance, and development there would protect better quality agricultural land elsewhere. The Cheshire studies set out to examine whether reclamation was possible and how the Ince Banks site compared to other alternatives in terms of cost and likely environmental impact. This question turned out to be one of the most interesting aspects of the study, revolving around three key issues: economic viability, ecological value, and hydrology, all of which required specialist technical investigation.

Economic Feasibility.   Examination of costs and feasibility quickly showed that there would be few technical problems in actually reclaiming the site from the estuary for building. The question of cost (or more properly economic viability) was much more difficult, and an engineering consultant was commissioned to look into this. The consultant's study sought to identify the methods by which reclamation could be carried out and estimated the cost of the job; these included a number of test drillings to assess the ground conditions to the north of the Manchester Ship Canal, and a study of servicing and development costs, to produce a broad-brush comparison of the cost of developing each of the sites.

Ecological Value.   The ecological report showed that the vegetated section of the banks had increased in size rapidly during the 1970s but the banks themselves had probably reached their maximum extent and had begun to erode due to movements of the channels in the Mersey. Bird populations also increased rapidly during the 1970s, and the banks formed the main roosting areas for wintering wildfowl and waders, accommodating 90 per cent of the estuary's bird population. However, the main feeding areas were found near the south bank of the estuary and would not have been affected by development. Because of the increasing ecological importance the area was made an SSSI (Site of Special Scientific Interest) in 1976, the SSSI site boundaries including Frodsham Marsh and a large part of the Mersey Estuary as well as Ince and Stanlow Banks.

Hydrological Effects.   The possible hydrological effects of reclamation were raised during consultations on the brief for the Mersey Marshes Study when the Port of Liverpool Authority and the Mersey Conservator objected strongly to any reclamation, arguing that it could reduce the tidal volume of the estuary and could affect natural scour, causing the existing navigable channels (which already require extensive dredging) to silt up. The Manchester Ship Canal Company was also interested in the reclamation of Ince Banks as a potential deposit ground for ship canal dredgings because their existing dumping facilities on Frodsham Marshes were almost exhausted. Mutual interest made it possible to agree on the issues to put to the Hydraulics Research Station. There were three main questions:

1. Would a small-scale reclamation above mean high water level adversely affect the regime of the river?

2. What would be the maximum reclamation without dramatically altering the hydrodynamics of the river?

3. What would be the maximum reclamation possible without adversely affecting other users of the river? (This assumes that the whole pattern of river flow might be altered but that the reclamation scheme could be so designed hydrodynamically that the flows though the main channels would not be adversely affected.)

The Hydraulics Research Station were asked about the cost and time which might be involved in answering these questions. Their response came in August 1977; it gave a reserved 'yes' to question one and suggested that the second question could be answered relatively easily, but that the third question could only be answered by testing with a large physical model. Construction and testing would cost between £150,000 and £200,000.

The Mersey Docks and Harbour Company were unhappy with even these limited findings, and challenged the Research Station over their views on the acceptability of reclaiming the smaller 600 acre area.

(v) The Future of the Villages

Figure 1 shows the extent of the present Stanlow complex and its containment within a series of natural boundaries formed by the Manchester Ship Canal, the M531, and the A5117, to the north, west, and south, and by the villages of Ince and Elton to the east. The first development beyond the 'natural boundaries' was the U.K.F. fertilizer plant built in the late 1960s; and Site 1 on the map is seen as a continuation of that decision, filling in the area between the existing plant and the ship canal but not extending further east onto Frodsham Marshes. The construction of the extension of the M56 motorway made another of the 'natural boundaries' of the existing complex, (the A5117), obsolete; taking away most of the through traffic and making it possible to expand the heavy chemicals industry onto the area cut off by the motorway.

The villages of Ince, Elton, Thornton-le-Moors, and Helsby were already affected to some extent by the existing complex, and Stoak could have been affected if the land between the A5117 and the M56 extension were developed.

The villages varied considerably in character. Helsby and Elton were the largest with populations of 4,200 and 2,400 respectively, the others being much smaller, with populations of around 200. Helsby is a long established village with a mix of housing ages and types including some very attractive residential areas on the wooded slopes of Helsby Hill.

Elton provides a sharp contrast. It was a small, run-down village in the 1950s but a private developer produced, and had approved, a village plan which was designed to increase dramatically the size of the village and improve its services and amenities. Consequently the vast majority of housing in the village was new estate-type development. It was the cheapest housing in the area and mostly occupied by first-time buyers.

Of the three smaller villages Ince and Thornton-le-Moors were both thought to merit designation as Conservation Areas, Ince also having some claim to archaeological and historic interest.

The Examination in Public of the County Structure Plan placed special emphasis on the problems of Elton and Helsby. The report made specific reference to the need to safeguard the amenity of these two villages by the use of buffer zones, mounding and planting schemes, and by control of the disposition of types of plant within the sites identified.

In the three smaller villages the situation was not so straightforward. The question of hazard (see above) had been examined but the Health and Safety Executive did not feel that buffer zones were justified in this area on hazard grounds, although safeguarding policies could still be justified on amenity considerations.

Ince and Thornton-le-Moors were undoubtedly affected by their proximity to industry. They suffered from smell, noise, flarestack operation and a high degree of visual intrusion from the petrochemical complex and from the two power stations at Ince. The study had to determine to what extent these environmental effects would be worsened by new development and whether adequate mitigation of further harmful effects could be obtained by the careful siting of plant and by relatively narrow (perhaps 400 metres) buffer zones associated with mounding and planting, or whether conditions would become so unpleasant that buffer zones of at least 1 km would be needed to protect amenity. Separation distances of that scale would severely limit the area available for industrial development on Sites 1 and 2. It might be considered against the public interest (in the widest sense) to restrict development in this way or to impose a cost penalty on firms by forcing them to develop on more expensive sites (e.g. Ince Banks) simply to protect the amenity of a few hundred people.

Another option would have been to clear some or all of these small villages to make way for industrial development. Such a policy had been put into practice on Teeside but this experience was not thought directly applicable to Cheshire's problem.

Changing External Circumstances

The adopted version of the Local Plan was produced late in 1985. During its preparation major changes took place in the national and international economic context which affected the local demand for industrial sites. Furthermore, important shifts occurred in the government's approach to planning.

The economic slump since 1979 had a considerable impact on the plan. The collapse of manufacturing industry in general, coupled with overcapacity problems in the petrochemicals sector and a substantial shift of investment to the United States, Europe, and newly developing countries conspired to produce a dramatic restructuring of the British chemical industry and a consequent substantial fall in the demand for new sites.

The Stanlow complex survived the spate of closures and rationalizations reasonably well, but employment levels have been cut back, and the smaller of the two oil refineries on the site (the Burmah refinery) was closed down. The much larger Shell refinery with its associated petrochemical and research functions is continuing to operate, but a linked Shell Chemicals complex at Carrington (30 miles to the east) which obtains its feedstocks from Stanlow has been severely hit by the recession, and substantial parts of the plant have been closed down, moth-balled, or sold off to other companies.

A limited revival has taken place since the trough of the recent manufacturing recession in 1983, and investment, productivity, and profitability have all recovered well at Stanlow and at the nearby ICI complex at Runcorn. However the industry has been moving into more specialized chemical products at the expense of its former concentration on large-scale production of bulk chemicals, and this, coupled with a move towards smaller and more automated chemical process technology, has meant that the recovery has not been reflected in a demand for large new sites. Two major new plants have been constructed (or are under construction) at Stanlow which between them represent over £150 millions worth of investment – a new oil blending plant completed in 1985, and a new catalytic cracker due for completion in 1988 – but both have been accommodated within the boundaries of the existing complex.

Government policies have also changed substantially since 1979 when work on the Local Plan began. In 1979 the National Economic Development Office was exhorting industry to exploit the benefits of Britain's new-found oil wealth by maximizing added value (i.e. producing chemicals from North Sea oil and gas liquid feedstocks). In 1983 this had changed to a much more modest strategy of encouraging development in the speciality chemicals sectors, and since 1983 the UK government has been virtually silent on the subject of industrial policy and has had little to say directly about the chemical and petro-chemical sectors (NEDO, 1983).

However another change in government policy could be of direct significance for those major complexes which, like Stanlow, are located in Special Development Areas. These installations previously enjoyed a major locational advantage as grants of 22 per cent of the total capital cost were automatically available and could be supplemented by tax benefits and other measures. This was particularly valuable to petrochemical projects which are invariably capital intensive. In 1984 automatic Regional Development Grants were discontinued and a new 'cost per job' criterion was imposed. It was feared that this might affect the ability of the Stanlow complex to attract new projects and keep its technological base up-to-date, but, as the examples above demonstrate, investment has continued, and, for the present at least, the future of the complex seems secure.

The changes in the economic situation since 1979 has also had some impact on the supply of possible development land. The closure of the Burmah refinery left a vacant derelict site of 230 acres within the boundaries of the existing complex. The site is proving expensive to reclaim and develop, and is taking some time to bring forward, but it does offer the possibility of a medium-term alternative to the green field sites on the periphery of the complex.

While these external changes were taking place (and some would argue undermining the need for making a plan at all) other changes were occurring within the plan-making process which had the effect of generating considerable delay in the production of the Local Plan. The Report of Survey – which was nearing completion in 1979 – led to a number of conclusions which, in turn, were incorporated in an 'Issues Report' produced in 1980.

The possibility of development in the area to the north of the Manchester Ship Canal known as Ince Banks was dismissed in the Issues Report as the Report of Survey had shown it to be unachievable within the ten year timescale of the

Local Plan, and in any case to be of unproven viability. The idea of reclamation for eventual industrial development was then progressed in another form in a project known as the Mersey South Banks Study.

With the Mersey South Banks Study now separated off, progress continued on the Local Plan itself. The 'Issues Report' was taken to public participation and consultation in 1980 and proved to be very contentious. This, plus the lack of urgency over demand for sites, inevitably led to delay; the problem was compounded by the 1981 local government elections which gave Cheshire a hung council for the first time. As control swung back and forth between the different political groups the Local Plan became something of a political football causing more delay.

The extension of the M56 motorway has now been completed, taking most of the heavy traffic of the A5117 (which had previously formed the southern boundary of the complex) and defining the outer boundary of the Gowy Marshes site more clearly.

Among the other local changes which have occurred since 1979, two are worth reporting here. The most pressing of the housing development and safety issues has almost resolved itself. Low demand and consequent low prices made further new house building in the village of Elton uneconomic, and as a result the builders decided not to develop three of the original seven blocks of land allocated for housing development in the village plan. This decision averted the immediate problem of housing development encroaching on an area considered a possible hazard zone in the plan.

The problems of whether the existing villages of Ince and Thornton-le-Moors should be cleared, and the definition of appropriate buffer zones remain of course, and re-emerged as major issues at the Local Plan public inquiry.

The local air pollution problems identified in the Local Plan Report of Survey were partly overcome by a programme of expansion and reinvestment at the U.K.F. fertilizer plant (formerly Shellstar).

Meanwhile, the Mersey South Banks Study which was basically a technical investigation of the environmental and engineering problems arising from reclamation, was underway as a jointly funded study involving the County and District Councils, the DoE, and some commercial contributors. It took quite a long time to develop a specification for the studies and to obtain the necessary funding, but work on the study was completed in 1987. It included separate 'topic' studies dealing with hydraulic and ecological impacts and reclamation methods, and a large-scale physical model of the Mersey Estuary was constructed for some of the tests. The funding for this model was organized separately. By completion these studies will have cost approximately £250,000 (plus an additional £72,500 for the model) and have taken two and a half years.

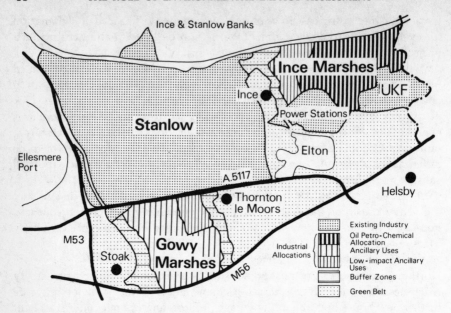

Figure 2. Land allocations in the Mersey South Banks Study.

The Local Plan

Draft proposals and policies were not produced until early 1983 (Cheshire County Council, 1983), the Plan being formally placed 'on deposit' later that year. Objections were considered at a local public inquiry in October 1984.

The land allocations and main policies of the plan were as follows:

LAND ALLOCATIONS FOR INDUSTRY

Expansion and redevelopment of existing firms to be encouraged within Local Plan area.

Land on the Ince and Gowy marshes is allocated for development as shown on the proposals map (figure 2). These allocations comprise:

INCE MARSHES:

410 acres for petrochemical and related development, 140 acres for ancillary uses with a limited environmental impact, and 75 acres for a landscaped buffer zone.

No road access to new development to be allowed through Ince Village.

GOWY MARSHES:

470 acres for ancillary uses, 82 acres for ancillary uses with limited environmental impact, 183 acres for a landscaped buffer zone.

Road access to new development only to be allowed from the A5117, and no development to take place until a surface water drainage scheme is implemented.

LAND ALLOCATIONS FOR HOUSING
New housing development to be limited to the replacement of existing dwellings, infilling, and the conversion of existing non-residential buildings where these are of historic or architectural significance.

ENVIRONMENTAL SAFEGUARDS
Local authorities to continue to monitor pollution levels (in liaison with statutory bodies responsible) and consult on all planning applications.

Village enhancement schemes are proposed in Ince, Thornton-le-Moors, and Stoak, and landscaped open areas will be maintained between the villages and any future industrial development.

The bulk transport of raw materials will be by rail, water, or pipeline whenever possible. All planning applications for major new plant must include an EIA.

GREEN BELT
A green belt is designated to cover most of the unallocated open land.

Its boundaries are shown on the proposals map.

INCE AND STANLOW BANKS
This area is to remain unallocated pending the outcome of further studies.

The Public Inquiry and Inspector's Findings

Since a large number of objections were lodged, the County Council was under a statutory obligation to hold a Public Local Inquiry which duly took place in October 1984. After considering the objections received, the Planning Inspector appointed to deal with the case decided that the main issues should be 'the demand/need for allocations', 'the safeguarding of the villages' and 'the Banks'.

Considering the heat generated earlier the inquiry was a curiously passionless affair. The opposition residents' groups, which had come together in 1980 under the umbrella title of the 'Mersey Marshes Study Group', presented a joint case to the inquiry, ably led by a local retired manager (ex-manufacturing industry). Their objection was based on the contention that there was no need for further land allocations in the short term, that the derelict Burmah site could meet any medium-term need, and that allocations against long-term need should be held over until the outcome of the investigations of 'the Banks' is known. They also maintained that the safeguarding measures proposed for the villages were inadequate.

In the present economic climate the case for major allocations was difficult to defend, and was not helped by a lack-lustre performance at the inquiry by representatives of the firms concerned who could not specify what the land would be used for or when it might be required.

The inquiry ran for nine days. The County Council received the inspector's recommendations in February 1985. His recommendations included a number of major alterations to the Local Plan's policies and proposals. The main change was the suggested deletion of the Gowy Marshes site from the areas designated for industrial use. Other changes included the recommendation to delete the 'ancillary' classification used to define areas for secondary uses on Ince Marshes. The inspector thought this area should remain in agricultural use, increasing the area of open land between the village and possible new industrial development. Also, and interestingly from the point of view of this book, the inspector recommended the replacement of the package of environmental safeguarding policies in the plan with a single requirement that planning applications for major development projects must be accompanied by an EIA.

The Lessons of EIA

Cheshire made no conscious decision to undertake an environmental impact assessment as such; rather, as the needs of the investigation were identified, it made sense to adopt appropriate EIA methodologies and approaches. Moreover, a decision was made to embark on a forward planning exercise rather than awaiting a development proposal. This decision was based upon a judgement that any further delay would be against the public interest and the interest of the firms involved.

The results were rather mixed, successes being offset by some disappointments. The successes included a clearer grasp of the issues involved and of the nature of the bargain to be struck between economic development and environmental protection in the lower Mersey Valley. A great deal of factual information was collected, many of the questions raised in the brief were answered, and a better understanding of the options available and the possible consequences emerged. However, the experience of using EIA raised a number of problems: problems of timescale; techniques of assessment; study management; and cost.

(a) Timescale

The major disappointment was the slippage in work programmes. The study was originally programmed to take eighteen months to complete, a timescale which had to be extended to three years. There were many reasons for delay: in part, complicated management arrangements and working systems; in part, underestimation of the scale and complexity of the task at the outset, as well as the unexpected shortage of information available. Progress was also slowed by the fact that the studies unearthed a number of technical problems such as those related to reclamation, air pollution and hazard; methods of dealing with those issues had to be found.

Another contributory factor was the difficulty experienced in working on new technical problems with specialist consultees. It took longer than expected to build up relationships of understanding between professionals. Some of the initial suspicion was overcome slowly while the problems of communication caused by different ways of thinking proved to be a major stumbling block. The apparent failure of experts to address simple questions added to frustration and delay. As the study progressed signs emerged that the environmental scientists were coming to terms with the needs of EIA. A growing force of consultants within universities began offering specialist advice in the environmental field. The statutory undertakers also became more aware of the special needs of assessment work; a good illustration is the Health and Safety Executive whose Hazards Advisory Group was set up for this purpose.

Rapid and unexpected change in external circumstances and a consequent loss of interest in new sites added to the delay. The study provided an object lesson in the risks inherent in policy-led forward planning – where the problem being addressed can change before the plan is produced.

The administrative approach to the issue also affected progress. Hindsight suggests that in moving from a 'special study' to a formal Local Plan the exercise became a sort of half-way house and may have fallen between two stools, struggling for sufficient technical credibility to be accepted as independent by the public and saddled with an extended and inappropriate consultation and participation process. With entrenched views on both sides the consensus (or best compromise) which was sought could not be achieved, and even after repeated attempts at consultation many of the major issues had not been satisfactorily resolved when the plan went to public inquiry.

(b) Techniques

Perhaps because Cheshire was one of the first in the field to use EIA in forward planning, it was not surprising that problems would arise in the technical aspects of the study. The most successful areas of investigation, like the air pollution survey, were those directly controlled by the EIA team, supplemented where necessary with bought-in technical expertise; the least successful were those dependent on statutory consultees and their ability to find time to produce detailed assessments in addition to an already busy office schedule.

Cases arose of experts either lacking the information required to answer the questions posed by the environmental impact study or holding onto the information because it was confidential. Examples of the former included health, pollution and ecological data. The main instance of the latter was hazard. Although data frequently existed it was both confidential and highly technical. The difficulty of dealing with hazard was aggravated by the Health and Safety Inspectorate which held most data and appeared to be overstretched by its everyday activities so that requests for additional information or analysis were seen as an unwelcome additional burden.

Health risk and pollution impacts on wildlife or vegetation were so difficult to measure that it seemed impossible, given the level of information available, to provide an adequate assessment. Nor was it simple to calculate the potential additional impact of each new chemical plant on the overall pollution levels found in the area. In a few cases the lack of adequate data proved to be an intractable problem. Other aspects of the assessment were effectively abandoned because of the prohibitive cost of seeking expert advice.

(c) Cost

The question of who should pay for EIA has been one of the central issues in the arguments about the practicality of assessment procedures in the planning system. While the developer should in principle be responsible, there are cases – such as the Mersey Local Plan – where general land-use policies and proposals must be agreed before individual proposals can properly be considered and which themselves require investigations of EIA in depth. There will also be cases where a developer's appraisal hinges on complex technical assessment and where the receiving authority feels it must commission an independent appraisal.

The Cheshire experience provides example of the varied kinds of investigation necessary to undertake an EIA and how expensive EIAs can be. For example, a hazard assessment of the type produced for Canvey Island cost over £400,000. An air pollution model for the lower Mersey valley capable of predicting the impact of new emissions on the existing pattern of pollution concentrations (a model similar to that developed by the Warren Spring Laboratory for the Forth Valley Study) would have cost an estimated £250,000 plus a local authority input of about two man-years.

Cheshire was only able to pursue those technical investigations necessary through a 'creative' approach to funding, using their own manpower and expertise as far as possible and finding partners with parallel interests to share costs in other cases. A great deal can be achieved in this way but finding the money can be as difficult and time consuming as carrying out the studies. Moreover, whether or not an assessment is carried out at all can be influenced by marketability rather than the need for research. The problem of financial resources is likely to become more acute as resource-starved and rate-capped local authorities find they do not have spare funds to commit to environmental impact studies.

Conclusions

Was the Cheshire environmental impact assessment necessary? Did the production of the Mersey Marshes Local Plan perform any useful function? Would a statutorily required EIA have overcome some of the problems encountered?

Some of the urgent issues which generated the original need for the Mersey Marshes Local Plan, for example, the housing versus industry conflict at Elton or the pollution problems at U.K.F. would have resolved themselves without the controlling policies in the plan. Other issues stemming from the inevitable conflict and competition for space between housing, heavy industry and other environmental considerations still have to be tackled. The Local Plan will perform a useful function by setting the overall land-use context for the next ten to twenty years and thus avoiding the repetitions of past poor and uncoordinated decisions and the cumulative build-up of further land-use conflicts. The removal of lingering uncertainty produced by the final adopted plan will also be much welcomed by local residents.

The main issues raised at the Public Local Inquiry, such as alternative siting possibilities and environmental protection, lay outside the normal scope of planning controls. This raises the question of the limit of planning powers and demonstrates the increasing public pressure to broaden the issues for consideration (the same pressure has been demonstrated at recent motorway, power station, airport, and nuclear waste dumping inquiries). In this context, the decision to make the Mersey Marshes Study very wide-ranging and to show how EIA could supplement the existing planning system by encompassing broad issues has been vindicated. Environmental impact assessment allows a more considered and thoroughly researched judgement than does the existing development control system.

The Cheshire experience suggests that the formal introduction of legislation on EIA could be valuable within the British planning system. But if EIA is to be useful some room for experimentation and development of new and better approaches will need to be possible within the legislation. Too little is known about the practical problems of carrying out studies of this kind to lay down fixed recommendations on the approach which should be adopted. Improvements are needed in basic data and in the analytical techniques used as well as changes in the approach adopted by some of the statutory bodies before really satisfactory results can be obtained from EIA. Many of these improvements will be best achieved through practice and experience.

A statutory assessment, on the lines suggested by Thirlwall and Catlow, with a full-time multi-disciplinary team to do the work, could have had some advantages for Cheshire. Day to day working relations between the various specialists within a team, all working to a common purpose, might have helped overcome misunderstanding between disciplines and improved working methods. The involvement of industry, local groups and local statutory bodies might have been more easily achieved. But some problems would not have been so easily overcome. A statutory system would have faced similar problems of patchy data and poor analysis techniques. The difficulty of getting to grips with issues like hazard and the impact of pollution on health and ecology would not have been any less. In addition, the imposition of a one year timescale for a formal EIA would have left little time for any original empirical work (such as that undertaken for the air pollution survey); indeed, a tight schedule could prevent the exploration of new methods. 'Difficult' alternatives, such as the possibility of reclamation, might have been overlooked.

Restricting environmental assessment procedures to new development proposals only will not be the most appropriate way of dealing with the sort of problems faced in the Local Plan example, where the conflicts arose from historic patterns of growth and change in industrial and settlement patterns. The impact of gradual change requires EIA treatment at least as much as single new large-scale proposals on green field sites. A hierarchy of EIAs may be the most practical solution. The higher level will set the general policy for an area, with lower level EIAs – perhaps equivalent to outline and detailed planning permissions in the existing system – being conducted for specific schemes or sites when actual proposals for these come forward.

EIA was in its infancy in Britain during the late 1970s. When Cheshire started to use assessment concepts and techniques they hoped for an unbiased resolution to the economic, environmental and political conflicts they faced in the Mersey Valley. Looking back, the expectation that even a well-executed technical appraisal could produce solutions favoured by all parties seems naive. EIA is no panacea of itself since the problems of the development process are deeply ingrained and reflect a community divided by sectional interests which are unlikely to be resolved by technical arguments.

The introduction of EIA in some form is part of a move towards the more open system of decision making being demanded by an increasingly environmentally conscious public. Ironically, the widespread adoption of EIA could prove something of a two-edged sword. There is a danger that EIA will be demanded for everything – each case will become a special case for the people concerned. This should be resisted. EIA is a useful supplement for special cases, but the time, effort and cost involved for both developer and local authority are substantial and can only be justified where development proposals involve contentious issues with a high technical content. The existing planning control system is well proved and has the ability to cope with the majority of development applications. To extend the use of EIA into lesser issues will simply devalue the techniques and undermine the existing tried and tested methods of planning practice.

REFERENCES

Catlow, J. and Thirlwall, C.G. (1977) *Environmental Impact Analysis*. Department of Environment, Research Report, No. 11. London: HMSO.

Cheshire County Council (1982) *Chemicals North West*. Chester: Cheshire County Council.

Cheshire County Council (1983) *Mersey Marshes Local Plan*. Chester: Cheshire County Council.

Clark, B. D., Chapman, K., Bisset, R. and Wathern, P. (1976) *Assessment of Major Industrial Applications: A Manual*. Department of Environment Research Report, No. 13. London: HMSO.

National Economic Development Office (1977) *UK Chemicals, 1975–85*. London: NEDO.

National Economic Development Office (1983) *Chemicals Industry*. London: NEDO.

APPENDIX

The background survey reports for the Mersey Marshes Local Plan were published by Cheshire County Council in 1980. The Report of Survey consisted of three main reports and several supplementary technical reports. All three reports concentrate on describing the existing situation.

1. Stanlow: The Existing Oil and Petrochemical Complex

The first report refers to the role of Shell's Stanlow operations in a national and regional context, and identifies its importance for the local economy, its future prospects and likely future land needs.

2. Environmental Conditions in the Mersey Marshes Area

The second report is essentially a baseline study. As in the first report, a great deal of original work is included, some in considerable depth. The report was not totally successful. The best sections deal with noise and air quality, where direct surveys have been undertaken to augment existing published data and so produce a more detailed picture of local problems. Other sections relied upon published reports and the policies of existing organisations and consequently were short of hard data relating specifically to the local area.

3. Physical Resources in the Mersey Marshes Area

The final report dealt with the physical resources of the area – geology, agriculture, ecology, archaeological and historic value. It attempted to provide a visual appraisal and assessment of conditions in existing settlements *vis-à-vis* future development. Although more successful than the second report, the subjective nature of some assessments and the difficulty of comparing unlike factors presented several problems.

Chapter Three

The Use of Environmental Impact Analysis in the Vale of Belvoir Coalfield

KEITH WILLIAMS

The EC Directive on environmental impact assessment raises the important question of whether the British planning system is capable of handling new formal procedures. In the changing context it is instructive to draw lessons from a British example in which the developer has produced an Environmental Impact Assessment (EIA), and to examine in some detail the technical, procedural and political issues involved in the application of EIA. This chapter describes the analysis carried out in the late 1970s by consultants to the then National Coal Board (now British Coal) in connection with their planning application to mine coal in north-east Leicestershire. EIA is evaluated at each stage in the decision process, beginning with the planning application and leading on to the subsequent public inquiry.

The Context

In 1977 the National Coal Board (NCB) announced that it had discovered the largest area of new coal reserves in Western Europe in the Vale of Belvoir in north-east Leicestershire, underlying 90 square miles (234 square kilometres) of predominantly agricultural land between Melton Mowbray, Nottingham and Grantham in the East Midlands region. As described elsewhere (Williams, 1978; Mann, 1981) the Belvoir coal deposits lie under a largely unspoilt rural area with attractive small settlements within commuting range of Nottingham, Leicester, Loughborough or Melton Mowbray (see figure 1). The Vale of Belvoir is a broad, low-lying, featureless basin of lower lias clay, largely consisting of pasture land for livestock farming, especially dairying which forms the basis for stilton cheese production. This area, which is famous for fox-hunting, is

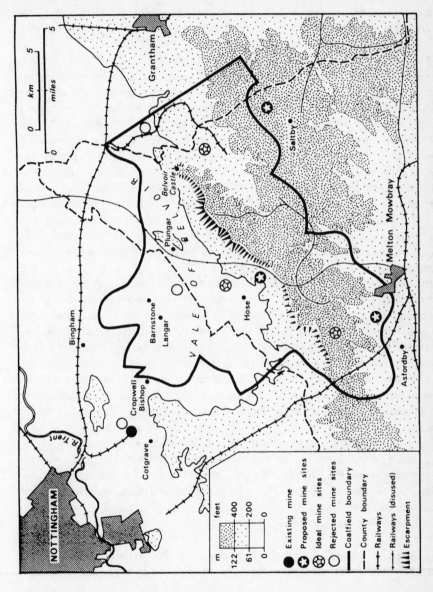

Figure 1. The Vale of Belvoir Coalfield proposals. (*Source: East Midland Geographer,* 7 (December), 1978)

dominated by the adjoining Jurassic uplands whose north-western edge forms a prominent, well wooded escarpment overlooking the Vale.

Construction work to develop the Belvoir reserves did not commence until August 1984 following a lengthy history of planning decisions which involved local authorities, environmental pressure groups and local objectors in a series of complex negotiations, the outcome of which was to reduce considerably the scope of the developer's original proposals.

The story began in August 1978 with the NCB submitting planning applications to three county councils (Leicestershire, Nottinghamshire and Lincolnshire) and three district councils (Rushcliffe, Kesteven and Melton Mowbray). Three coal mines and related tips or surface works were proposed at Hose, Asfordby and Saltby in Melton Mowbray District (see figure 1). The Secretary of State 'called in' these applications in January 1979 so that he could make the final decision after a Public Inquiry had been held. A preliminary meeting was held in May 1979 to identify the main issues which were to be the subject of the Inquiry and to provide opportunity for additional relevant information to be requested. The Public Inquiry, under the chairmanship of Michael Mann, QC assisted by two expert assessors, lasted for eighty-four sitting days between October 30, 1979, and May 2, 1980. In November, 1980, the Inspector's Report (Mann, 1981) was submitted to the Secretary of State, who announced his decision on March 25, 1982.

Before submitting its planning application, the NCB had spent a year considering a report from its consultants on surface works (Leonard and Partners, 1977), which contained a detailed environmental impact assessment. Indeed, before the full extent of the coal reserves had been revealed, the NCB had published an earlier report (Leonard and Partners, 1976) which had also incorporated an environmental impact assessment.

At about the same time as the NCB was producing its environmental impact assessment, a research report on environmental impact analysis commissioned from consultants by the Department of the Environment was also published (Catlow and Thirlwall, 1977). This report concluded that a system of environmental impact analysis was needed in the United Kingdom for certain large-scale and complex development projects. Thirlwall outlined five important criteria (Thirlwall, 1978) and it would be useful to summarize them briefly before discussing how far the NCB consultants' EIA conformed to these recommendations.

Thirlwall recommended that:

1. EIA should be used at an early stage as a design tool rather than as a test of a proposal, and it should allow scope for an effective examination of realistic alternatives.

2. Analysis should include social and economic impacts as well as those affecting the physical environment.

3. An impartial report should be prepared which avoided making recommendations about planning decisions, and which excluded subjective judgements about the relative importance of the different impacts. Facts and opinions should be clearly distinguished, and opinions should be supported with a qualitative analysis and reasoned explanations of the opinion. Although environmental impacts are in large part unquantifiable, simple indicators of scale, such as the number of people affected, etc, would be useful, providing such figures are accompanied by a careful assessment of the environmental change experienced. A ranking order could be used to express the performance of each alternative against each impact studied, although there are dangers if this judgement is made on crude numbers alone. A further danger lies in the natural weakness for simplication leading the decision-makers to assume all impacts are of equal importance and simply adding up the relative rank orders or scores.

4. The planning authority and developer should co-operate in setting up and supervising the study, perhaps with a steering committee to guide the experts drawn from a wide range of disciplines.

5. EIA would be applied to comparatively few projects and could be integrated into the existing development control system

(a) by inviting early notification to the planning authority of an impending development;

(b) a county planning authority would normally be responsible for deciding whether an EIA was required;

(c) the responsible authority should determine what environmental impacts were relevant to a particular planning decision and should be included in the analysis;

(d) the alternatives to be considered should be decided jointly by the planning authority and developer since any alternative unacceptable to the developer would not be a realistic one;

(e) there should be provision for public participation at a time when the decision was being made, plus public consultation and comment on the EIA before a planning decision was taken;

(f) external costs of the analysis should normally be shared equally between the developer and the planning authority.

Before considering how closely the NCB consultants' environmental impact assessment conformed to these Catlow and Thirlwall guidelines, however, it would first be appropriate to outline briefly the developers' environmental impact assessment procedures.

The Consultants' Environmental Impact Assessment

The consultants' evaluation (Leonard and Partners, 1977) was published in a voluminous 700-page document which describes the complex exercise to reconcile technical mining considerations, the costs of surface works and the environmental impacts. As already outlined elsewhere in a more extensive critique of the NCB consultants' work (Williams, Hills and Cope, 1978), their appraisal or plan-evaluation involved four main stages:

(*a*) Sieve map analysis to identify six possible mining sites.

(*b*) Impact analysis using four quantitative indices plus the surface costs of nine mine site options.

(*c*) A similar analysis with six quantitative indices for six site strategies, or combinations of options.

(*d*) A comparison with underground mining considerations.

A. Sieve Map Analysis

The well-established planning technique of sieve map analysis was used to eliminate areas unsuitable for mines or dirt tips. Eleven criteria were mapped and then generalized to ¼ km grid squares. Maps were prepared for:

1. Rail accessibility: areas within 2½km of existing or recently-closed tracks with no severe gradients or curves.

2. Road accessibility: a simple gravity formulation relating settlement capacity, spare capacity of roads and journey times from Bingham, Grantham and Melton Mowbray.

3. Public utility services: power lines, oil and gas pipelines which would have to be diverted.

4. Slope analysis: slopes steeper than 1:25 which would require costly ground preparation works.

5. Vegetation and wildlife habitats: areas with dense woodland, mature hedgerows and undeveloped land.

6. High agricultural potential: land classified as grades 1 and 2, plus the best-drained grade 3 land.

7. Low soil capability: wet, shallow or steep soils which would increase site-preparation costs and inhibit the restoration of dirt tips.

8. Proximity of settlements to mine buildings: zones of 1½ km around population centres.

9. Proximity of settlements to tips: ½ km zones around population centres.

10. Landscape value: scores assigned to various elements of landscape.

11. Statutory designations: zones of ½ km around conservation areas, sites of special scientific interest, and sites of geological, historical and archaeological interest.

These maps were then superimposed on each other to indicate both 'no-go areas' and sites close enough to the ideal underground locations for mines, which were large enough for the construction of mines and dirt tips and which would not result in excessive environmental damage. The use of the sieve maps led to the identification of six possible mine sites: Hose (near the scenic escarpment); Langar (near a rural industrial estate on a former airfield); Cotgrave (close to an existing modern coal mine) in the centre of the coal reserves; Asfordby (next to a small steelworks) in the south; and in the east at Saltby (on a disused airfield); and at Stenwith using an overhead conveyor belt to a tip at Normanton (another disused airfield). None of the sites considered was in an ideal location according to underground technical mining conditions. Equal weighting of all relevent maps, however, failed to eliminate sites in environmentally attractive and sensitive areas, particularly at Hose, which subsequently led to great controversy and provided a focus for local objectors.

B. Impact Analysis of Nine Mining Options

Various combinations of total or satellite mines, service shafts, mineral outlets, coal preparation plants or tips were next investigated. Nine different groups of mining operations, or options, were considered as alternative strategies. For each mine or tip site, four quantitative indices were calculated for:
1. the visual intrusion effects;
2. the noise impact;
3. landscape loss;
4. agricultural loss.

The worst case in the study region was noted and then each option was scored as a percentage of this case; in most, but not all measures, the worst case was one of the sites under consideration. A comparison was also made of the relative surface costs of each option. A matrix was prepared to indicate the relative performance of each option against the four selected environmental impacts and their surface costs. Impacts and costs were assumed to be of equal weight and importance, and the procedure used to evaluate performance consisted of classifying each impact in terms of whether it was above average, average, or below average for the range of options as a whole.

1. Visual Impact.
The visual impact measure represented a simple summation of five different calculations based on the existing residential population, the landscape quality, and local road users. Using the visual intrusion of the mine winding towers at the existing modern Cotgrave mine, the zone of visual

intrusion was defined as a 3 km radius for each site. The population able to see the top of the winding tower or coal preparation plant was expressed as a proportion of the worst case in which the largest rural population within a 3 km zone could see the tower within the study region. Landscape quality was assessed by scoring elements in the landscape such as topographic variation, woods, etc, so that the area in the two highest categories could be expressed as a proportion of the worst case in which the entire 3 km zone was in the highest landscape quality category, even though no mine was suggested in such a scenic area. This score was added to another score derived from the number of conservation and special scientific areas within visual range expressed as a proportion of the largest number of such areas found within a 3 km zone in the study region. Finally the lengths of road within visual range of the mine tower were multiplied by an ordinal measure, the traffic flow class. The resulting three scores were weighted equally and a simple average score derived for each of the nine options.

It is not clear how the population figures were obtained in order to calculate some of these measures. The assessment of landscape quality tends to be rather subjective even if dressed up in a cloak of apparent objectivity through assigning scores for specific features. The measures were obviously not all weighted equally in reality: a car driver is less likely to be troubled by the sight of a winding tower than a rambler or someone choosing to live somewhere because of an unspoilt view.

2. Noise Impact.   Potential noise contours within the same 3 km radius of a mine site were estimated for the construction period, while three different noise contours were calculated for the production phase, depending on whether the source of the noise intrusion was from the mine site or from a railway, or by day or by night. In each of the four calculations, the resident population was expressed as a proportion of the maximum possible population found in the study region within a 3 km zone. The three production noise scores were weighted equally to give a new score which was weighted at three times that of the construction noise score. The resulting construction and production scores were added together and averaged to produce the final noise impact score for each of the nine options.

3. Landscape Impact.   Landscape quality was subjectively assessed using a method similar to that employed in the large, more diverse area covered in the West Midlands Regional Study. Again, the 3 km radius impact area was used to take into account the total area from which mine buildings could be seen. After weighting the highest category of landscape quality by a factor of three, and the second category was multiplied by two, the resulting scores were expressed as a proportion of the worst possible case in which the entire 3 km zone was classified as grade 1 landscape quality.

Although slightly different since a weighting system has been incorporated, this landscape impact appeared to largely duplicate part of the calculations used in scoring the visual impact. No attempt was made, however, to consider the landscape impact of dirt tips since it was argued that these tips would change in nature during the life of a mine. Furthermore, the areas of high landscape quality shown on the sieve map of landscape value appear to have been obscured from the 3 km zone around a mine site.

4. Agricultural Impact. The estimation of the loss to agriculture over a 75-year period required a long series of calculations, each involving assumptions about the areas involved for dirt tips and surface works, the likely future agricultural productivity, and the returns per acre, together with the ease of restoration. A 50 year mining period was assumed with no restoration of the mine site itself, but the tip was assumed to be returned to full production over a 20 year period, with the estimated returns associated with the optimum rotation system for each site being calculated. Net average crop yields per hectare were used to indicate the total value added over the 75 year period in the absence of a mine. The net value added during the restoration period was estimated by subtracting assumed overhead costs for different crops at each site from the estimated gross value of production lost during the restoration period. To produce monetary values for the total net loss, the estimated value added figure was deducted from the assumed value added with no mining.

The impact of mining on the agricultural conditions in the immediately affected farms was not included in the analysis. The forecasting methods used tended to demonstrate how advantageous it would be to return partly derelict sites, like the former airfields at Langar, Saltby and Normanton, to agricultural production following restoration works. The NCB consultants appeared to be optimistic about the ease with which the dirt tips could be returned to arable uses although many of the tips elsewhere currently undergoing restoration appear to have been used for grazing only.

C. Evaluation of Six Strategies

This impact analysis, however, was constrained by mining factors which necessitated a mine site in each geographical division of the coalfield – south, central and east – with the largest mine in the more productive central area. The consultants argued that sites could only be compared with others in the same geographical zone. Since Asfordby was the only option proposed for the southern part of the field, it could not be compared with an alternative during the evaluation. Indeed, its impact relative to dissimilar sites in the central and eastern areas may have altered both the calculated impacts in relation to the worst case and the rankings of other sites. Only two options, Saltby and Stenwith/Normanton, were considered in the east, and since Saltby consistently out-performed the latter, Stenwith/Normanton was therefore dropped from further analysis. Mines at either Langar or at Hose were the preferred options to result from the impact analysis stage for the central area.

The third stage of the NCB consultants' analysis consisted of the formulation and evaluation of six mining strategies, based on the options already considered but reflecting the mine distribution constraint. Site options were therefore combined into groups of three with a possible site in each zone, south, central and east. The evaluation then simply aggregated the scores of the options which had been grouped together into these strategies, but added two new additional criteria relating to traffic effects, both for the impact on road capacities and for the traffic instrusion impact.

For the purposes of the exercise it was assumed that all miners would live only at Melton, Grantham or Bingham, and would work in the nearest mine, travelling on roads giving the shortest journey-to-work. It was arbitrarily assumed that 2000 houses would be located in the Melton Mowbray area, 1000 in the Grantham area, and 500 houses required at Bingham. Computations on traffic flows indicated the traffic overload on the existing road system if the network's theoretical capacity or the flow projection for 1995 was exceeded. The overloaded mileage of the roads was added to the total number of commuter vehicle miles on the system to produce the impact scores. The traffic intrusion impact was estimated for the construction traffic and for the total traffic and the heavy vehicle traffic during the production phase. The construction traffic impact over the first six years at each mine was measured as the sum of the village populations subject to a 50 per cent increase in heavy vehicle traffic and expressed as a percentage of all village populations affected by this traffic. For the production traffic, the population of villages subject to a 25 per cent increase in estimated total traffic for 1995, plus the population of villages with a 50 per cent increase in heavy vehicle traffic due to servicing the mines, was expressed as a percentage of total population affected. A combined score was then produced in which the production intrusion was weighted as nine times that of the construction traffic intrusion.

Strategies were thus evaluated on the basis of the six environmental factors, whose measurement raised many doubts about their validity, together with the surface capital costs of the engineering, architectural and landscaping requirements at each mine. All scores were adjusted to a common scale in which the lowest score was set at 10 per cent and the highest at 90 per cent, with the other scores proportional between these two figures. An attempt was also made to weight the population-related impact scores according to the consultants' perceived relative importance of each factor: a weight of 17 each was assigned for visual and landscape impacts, 15 for agriculture, 12 for noise, and 7 each for traffic intrusion and for road capacity, contrasting with the weighting of 25 for surface costs. It was claimed that no significant alterations in rankings emerged from the use of weighted and unweighted scores.

The overall result of this evaluation by the consultants, which still excluded underground factors, was the ranking of the possible strategies for development of the coalfield. It was noted that all strategies would have a considerable impact on the environment, but that no single strategy appeared overwhelmingly superior on all environmental grounds or had consistently better results. The preferred strategy would have mines at Langar, Asfordby and Saltby, while the next choice would have total mines (with winding towers, coal preparation plants and tips) at Hose, Asfordby and Saltby.

D. Comparison with Underground Mining Factors

Having identified an environmentally more acceptable strategy, this 'preferred' surface strategy was then ruled out as failing to meet certain mining requirements since the Langar mine would have been more expensive to operate than the Hose mine. As a result, the NCB's planning application was for a mine at Hose to produce 3 million tonnes of coal per year, a 2.2 million tonne mine at Asfordby, and a 2 million tonne mine at Saltby to be started four years after the other two mines.

The Lessons of the Developer's EIA

How closely does the NCB consultants' environmental impact assessment conform to the five criteria outlined by Catlow and Thirlwall?

1. As a Design Guide

At first sight the study certainly appeared to outline how the sites were selected and alternatives evaluated, but closer inspection suggested that it may have been intended more as a test of a proposal, particularly for the most controversial central mine site which had previously been selected and evaluated in the earlier consultants' feasibility study before the full extent of the coal reserves had been determined. There is, however, little doubt that environmental considerations caused the relocation of all three mine sites away from the optimum underground mining locations. This appeared to have resulted from the initial sieve map analysis, though, rather than the environmental impact assessments.

2. Inclusion of Social and Economic Impacts

The NCB's specific exclusion from their consultants' brief of housing requirements, infrastructure, manpower and subsidence considerations destroyed much of the potential value of this exercise as a true EIA. All impacts and their geographical incidence could not be systematically identified. The resulting lack of comprehensiveness led to ignorance about many of the most significant potential impacts likely to affect the planning decision, especially those of strategic and regional importance with indirect impacts, or beneficial effects. As a result environment was narrowly interpreted as a limited range of physical impacts, generally only within a local or 3 km radius of each proposed mine site.

3. Impartiality and Objectivity

Although the consultants collected and mapped an invaluable baseline inventory of physical conditions in the study region, these facts were swamped by excessive quantification. Public debate and participation was inhibited by the provision of so much information in an unfamiliar format. Too many assumptions and simplistic generalizations were inevitably hidden within the mass of calculations.

Unquantifiable environmental impacts need simple indicators of scale accompanied by a careful qualitative description of likely environmental changes. By going too far and trying to be too sophisticated, the consultants' judgements were made on crude numbers alone. The trap of over-simplification led them to assume that all impacts were of equal importance. The scores used appeared to be precise but several unlike, equally-weighted measures were added to other indicators and, in the case of the strategies, to scores obtained for other mining options. On several occasions subjective weightings were attached to the scores without justification. The totalling of the number of times an option was ranked as above-average or below-average was also a dubious use of ordinal data. In fact, the consultants' earlier Vale of Belvoir Project Feasibility Report was perhaps more satisfactory since it did not push the quantification of measures so far. Ironically this earlier study appeared in many respects to be closer to a true EIA. One suspects that this entire study may have been regarded as a public relations exercise to demonstrate that the NCB was aware of environmental issues. It should perhaps be seen as a response to the criticisms raised over the Selby coalfield in North Yorkshire where NCB plans for a huge drift-mine with satellite shafts received planning permissions in 1976 and 1977.

4. Co-operation with Planning Authorities

Although the various planning authorities supplied data for the study, they do not appear to have been consulted about the NCB consultants' environmental impact assessment.

5. Integration with Existing Development Control System

As a result of the lack of contact with the planning authorities prior to the production of the NCB consultants' analysis, the local authorities were not aware of the impending development, and could not determine what environmental impacts were relevant or what alternatives should be considered. The consultants' preferred strategy was found to be an unacceptable alternative for the developer, the NCB. The early circulation of the detailed consultants' study, however, did promote public participation and discussion at no expense to the local authorities' ratepayers, although as taxpayers or energy-purchasers they may well have contributed.

Possible Changes with Local Authorities' Involvement

There is little doubt that the complexity of the exercise would have greatly increased if the local authorities had been fully consulted about the environmental impact assessment procedures at an early stage in the exercise. There would not have been just one planning authority involved, but three counties and three districts would have had to work together with the NCB and its consultants. Such a committee would have had to face problems over procedures, as well as difficulties over inter-corporate relations and officer-member relations. Each authority would have had its own views on priorities and interpretations, as well as on the alternatives to be considered. The greater experience with development control procedures would probably have meant that the excessive quantification and summing of scores would have been avoided and a more comprehensive view taken. Priority would probably have been given to producing background information and maps, supported by reasoned qualitative discussions to assist the decision-makers make their own judgements and attach their own weightings to the various factors presented in an ideal environmental impact assessment.

It is tempting to suggest that a final decision might have been reached by 1978, six years earlier than the eventual approval of the modified coalfield proposals had the NCB agreed to prepare an environmental impact assessment jointly with the planning authorities, even if that had taken two years. Such a scenario, however, would have required an agreed national energy plan to have avoided debates about the need for the Belvoir coal or the time when it would be in the national interest to develop the coalfield. If national needs for the coal could have been covered by a national planning forum, such as a Select Committee of the House of Commons or the Lords, then the environmental impact assessment could have concentrated on the protection of the local environment. At the time, various bodies wanted a two-stage exercise or a Planning Inquiry Commission which might have avoided mixing together discussions over national and local issues and concerns.

Local authority planners would almost certainly have been primarily interested in the social and economic impacts, even if only to ensure that the developer made a realistic financial contribution towards the direct and indirect consequences of the proposals and to avoid the additional burden from falling on the local authority rate-payers. In addition to all the difficulties created by the unknowns, uncertainties and subjective assessments associated with the physical impacts of the coalfield proposals, a whole new set of estimates and assumptions would have to be made about the impact on manpower and about the numbers and the distributions of houses which would be required. Little real knowledge would exist about the number of miners who would be prepared to commute long distances from less productive or worked-out pits, or about the willingness of local workers or school-leavers to move over to well-paid mining jobs. The numbers involved might have required the preparation of comprehensive local plans to take into account the existing distributions and spare capacities of a wide range of infrastructure and services. Local authority planners might also have been more interested in the benefits of construction sub-contracting, local expenditures on mining equipment and supplies, and the local consumer multiplier effects from additional workers during the construction and the production phases. They might perhaps have also been less concerned with landscape and visual effects than with the impact of subsidence on properties and on drainage systems, and might also have taken a shorter-term view of the impact on agriculture.

From the reactions of the planning authorities after the publication of the NCB consultants' report, however, it is possible to gauge how they might have reacted to an approach to guide the environmental impact assessment in the initial plan-preparation phase. In September, 1977, the North-East Leicestershire Coalfield Working Party was established, bringing together the six affected local authorities with the NCB and, when required, statutory undertakers and other public bodies, in order to examine and cost the infrastructure requirements and other selected factors likely to result from the development. Local authority participation did not involve accepting that the need for Belvoir coal had been proved, nor did it prejudice or pre-empt any subsequent decisions which the local authorities or the NCB might wish to make.

One subsidiary working group was established to investigate the infrastructure implications and costs for housing, transportation and ancilliary requirements like sewerage, water supply, shopping, education, ambulances, libraries and other social services for a limited range of alternative development options or assumed distributions of extra population. Even so, the three-mine, two-mine (Hose and Asfordby) or single-mine (Hose or Asfordby) options which were considered did not represent an exhaustive range of responses, as might be expected in an EIA, nor were the cost criteria comprehensive since additional revenue costs or those resulting from the indirect multiplier effects were excluded. Another working group focused on colliery dirt disposal, which included the comparative costs, safety, environmental or engineering feasibility of underground stowage and remote disposal methods, especially at the Bedfordshire brickpits. A third sub-committee dealt with subsidence and drainage issues.

The Interim Report of the Working Party was published in April 1978, and the Joint Working Party continued to meet to exchange basic information. Although local opposition groups complained about their exclusion, and some local authorities were still pressing the NCB to make more information available, the Working Party undoubtedly helped to foster a working relationship over contentious issues. Its work did counteract the deficiencies of the limited environmental impact assessment which had originally been presented. Despite its relative success, however, Leicestershire still felt obliged to produce its own total impact assessment, summarizing and commenting upon some of the physical environmental impacts from the NCB consultants' report, adding an indication of the likely capital costs and summarizing for elected members of the County Council the issues involved over dirt disposal, subsidence and drainage, but also extending an analysis of the likely multiplier effects and the jobs which would be required (Leicestershire County Council, 1979). This document was probably closer to a true environmental impact assessment than the consultants' document.

In another series of moves which demonstrated the planning system's ability to evolve to meet changing circumstances, both Leicestershire and Nottinghamshire County/Rushcliffe District held a series of well-attended public meetings to provide information and to assess local opinions and reactions. Local pressures to reach decisions about their attitudes to the planning applications were resisted until more information had been provided by the NCB and until after the local authority officers and elected councillors had heard the views and fears of the people affected. During the course of these meetings no-one wished to discuss any aspect of the consultants' environmental impact assessment, so it cannot be argued that the EIA encouraged public participation or discussion. This could have reflected the fact that the volumes were not circulated widely enough or that objectors did not have enough time or energy to try to absorb this work. On the other hand, the EIA could have been dismissed as being such a technical exercise that little merit was seen in arguing about the consultants' calculations, assumptions or weighting scheme. Most of the objections, however, did revolve around the direct physical environmental impacts, such as subsidence, drainage, noise and air pollution, increased traffic flows on narrow country lanes, loss of productive farmland, or the visual intrusion of mine and tip sites. A number of local people also pointed to the social and economic problems experienced in the 1960s with the nearby Cotgrave Colliery due to the rapid growth of the mine and of NCB housing to accommodate the sudden influx of miners from several distant coalfield areas. Concerns were also expressed about the financial burdens on local communities, together with the social and economic implications and disruptions to established social patterns.

Many of the procedural changes which were introduced as part of the planning system's response to the Belvoir Coalfield proposals can be traced back to the proposals of George Dobry in his review of the development control system (Dobry, 1975). Pre-decision discussions between local authorities and the developers, the various parties reaching an agreed statement of fact before any Inquiry, pre-Inquiry meetings to clarify issues or prepare an agenda, and public participation, have all been introduced in the Vale of Belvoir case. In particular, Dobry's suggestion that major applications should produce an impact study to back up their proposals seemed to have led to the consultants' pioneering attempt to use environmental impact assessment in such a complex and difficult situation – even if a suspicion remains that its real purpose was to fudge issues for the benefit of their client.

The Role of EIA during the Public Inquiry

On January 23, 1979, the Secretary of State for the Environment exercised his powers under Section 35 of the Town and Country Planning Act, 1971, to 'call in' or direct that the three applications be referred to him for decision because they involved a substantial departure from the development plans and raised issues of national and regional importance. In his Rule 6 statement in conformity with the Town and Country Planning (Inquiries Procedure) Rules, 1974, Peter Shore, the then Secretary of State for the Environment, said that he would take into account environmental factors, the extent to which coal would otherwise be economically available, and all other planning aspects of the proposed development. An inspector, Michael Mann QC, was appointed with two assessors to hold a local inquiry concerning the NCB's planning application.

The Public Inquiry was held at Stoke Rochford Hall, Grantham, from October 30, 1979, to May 2, 1980; a total of eighty-four sitting days plus an evening session at Hose and another at Coalville in North West Leicestershire. There were 456 written representations, evidence was given by 136 witnesses, representing some sixty organizations, and a further sixty-three individuals also gave evidence.

Unfortunately, the existence of the NCB's EIA did not shorten discussions or clarify issues at the inquiry. In the case of each of the key impacts identified at the inquiry, the developer's EIA was either challenged or superseded by alternative appraisals, as the following paragraphs demonstrate.

Traffic Impact

On behalf of the other local authorities, Nottinghamshire carried out its own road traffic noise predictions on a with-and-without mining basis to identify the worst affected locations and to specifiy the houses which would be affected by increased disturbance during the production phase. The county carried out a similar assessment of railway noise. Leicestershire also compared existing and predicted traffic flows for 1996 on a with-and-without mining basis; they were able to identify points in the network where improvements to traffic capacity were required, and during the course of the inquiry all except one of these improvements was agreed by the NCB.

The coalboard also offered to pay for the construction work needed to improve access to the mines. The 'Alliance' too, had prepared a 1996 commuter pattern for each mine, based on an arrival and departure survey at Cotgrave, in order to calculate routes subject to capacity pressure and where traffic intrusion would occur. Again they were not dependent on the consultants' traffic impact studies, preferring to incorporate the multiplier effects of development into their estimates of housing and its implications for traffic.

Noise Impact

Melton Borough Council did not accept the NCB's evaluation of mine site noise but carried out their own studies, drawing upon complaints of construction noise, especially from blasting, at Selby, North Yorkshire. Once again the NCB's impact assessments were wasted or duplicated, or at best provided a focus against which other specialists could react and make modifications.

Agricultural Impact

The Planning Inquiry again demonstrated the vulnerability of the initial impact assessment to changes in the assumptions. Leicestershire County Council forecast waste disposal tips for 75 years rather than for the 50 years used in the NCB's calculations. In addition, substantially more land would be required for off-site tree planting, while substantial areas of agricultural land would also be needed for housing, road widening, sewage works, rail links, re-routeing gas pipelines, underground electricity cables, etc. It was also pointed out that some of the land taken would represent a permanent loss to agriculture rather than the temporary losses assumed by the developers, other areas would be out of food production for long periods with no certainty that restoration would ever be fully successful. Even the return of Saltby airfield to agriculture after restoration would not represent the gains predicted since reclamation work has already started in advance of mining. The damage to land drainage through subsidence would be considerable but this factor had been neglected in the original impact studies.

Severance of farm units had also been under-estimated by the developers: from evidence at the inquiry, only two out of the nineteen farms, which would be affected by the mine sites, would have to be purchased as a whole, but the overwhelming majority would have to carry the same fixed costs (machinery, etc) over a smaller or less productive acreage. The 'Alliance' described the actual impact for each of the farms, indicating the stocking density, labour efficiency and gross margins, producing a more realistic picture than had been given earlier by using parish or regional averages. It was noted that

> the NCB's method of assessing agricultural impact has 4 major shortcomings (use of value added, failure to capitalise, disregard of secondary development and failure to quantify the effect of subsidence). NCB has never had before it an accurate assessment of the total agricultural impact which is much greater than has been admitted. (Mann, 1981, para D5.6.13, pp. 288–9)

Visual/Landscape Impacts

It was the consultants' landscape and visual impacts, however, which came in for the most scathing criticisms from local government and environmental interests. Leicestershire noted that

> The whole of the N.E. Leics. Prospect is an area of high landscape quality. This is the subjective judgement of most people who visit the area and does not require justification by psuedo-scientific methods of measurements. (Mann, 1981, para D1.5.9, p. 240)

The County argued that visual intrusion had to be assessed from points within a 5 km rather than 3 km radius of each site. They pointed to the ugly and offensive nature of spoil tips which had been ignored, as well as to other features like road improvements, new railways, electricity power lines, and the effects of subsidence which would all detract from the landscape. They were able to describe the disturbance to the historic fabric, and list sites of archaeological importance, national geological importance; and they also carried out a survey to reveal sites of ecological interest. Melton Borough Council bluntly dismisses the consultants' visual and landscape impact assessment, stating that

> The techniques used by the N.C.B. to assess landscape quality are not convincing. They are largely subjective and do not take into account public opinion on visual acceptability. (Mann, 1981, para D4.11.2, p. 271)

The Countryside Commission was also critical of the assessment techniques arguing that landscape quality could best be appreciated by describing landscape features not portrayed on an ordnance survey map and that large changes in scores were fixed by subjective choice to reflect only minor changes in reality. The 'Alliance' were very critical of the whole environmental impact assessment, especially the landscape quality aspects:

> Not only were there defects and inconsistencies, but each stage was deficient in scope, method and the interpretation of results. The selection process was wholly inadequate and the outcome wholly misleading. (Mann, 1981, para D5.8.12, p. 292)

It was also observed that

> the site selection process of NCB does not provide convincing proof that the proposed sites are the right or only sites. Underground and surface considerations have been brought together in an manner which precludes proper listing of an appropriate range of alternatives. There is no evidence to show that EIAs have been used to optimise mine locations. (Mann, 1981, para D5.8.24, p. 293)

The only other occasion during the Inquiry when EIA was mentioned was in the opening statement of the Town and County Planning Association: a development of this kind should be preceded by an EIA on the lines recommended in DoE Research Report 11 (Mann, 1981, para G6.3, p. 309). Further TCPA participation was precluded, however, due to the difficulty of funding participation, and it was suggested that the government should remedy the situation by establishing a method of funding bona fide objectors at major inquiries.

The Inspector's Recommendations

The Inquiry ended in May 1980 and the Inspector presented his report to the Secretary of State for the Environment on November 26, 1980. (Mann, 1981) On the need for Belvoir coal the Inspector stated that

> I am of the opinion upon the evidence that it is somewhat more likely than not that there will be a need for a supplement to indigenous deep-mine capacity at about the time [1995] NELP could become fully operational. (Mann, 1981, para 17.2.1, p. 100)

He also observed that this capacity in the 1990s 'can be met only by the exploitation of NELP' (Mann, 1981, para 17.3.1, p. 100). In his judgement it appeared that the NCB's arguments had been stronger.

> The 3 selected mine sites represent the best compromise between operational and environmental factors. The development of the 3 sites would be acceptable in visual terms and the operation of the sites would not have unacceptable consequences in terms of road and rail traffic, atmospheric pollution, dust, noise, vibration, or water pollution. (Mann, 1981, para 17.4.1, p. 100)

During the course of the Inquiry, the NCB had entered into various agreements with the local planning authorities and the water authorities, and also given assurances to various public bodies. Many of the conditions which the counties wanted to see imposed should planning permission be granted had also been agreed, while the Inspector gave his recommendation for conditions which were still in dispute. In a recommendation which pleased the objectors, however, he considered that the construction of tips at Hose and Saltby would in visual terms be totally unacceptable. He therefore recommended the refusal of permission for the construction of tips at Hose and Saltby, stating that

> the case on need is neither sufficiently definite in point of time nor strong enough in terms of quantity to warrant the environmental harm which would be caused by the two tips. (Mann, 1981, para 17.6.4, p. 100)

In this way the Inspector's recommendations, if accepted, would have forced the various parties to consider a compromise since neither the NCB not the objectors would have been completely satisfied. Some, but not all, of the conclusions of the NCB consultants' environmental impact assessment appeared to have been accepted by the Inspector, although whether as a result of the consultants' exercise or of the arguments presented at the Inquiry cannot be determined. It seems safe to conclude that the consultants' EIA acted as a catalyst and stimulated critical comments and additional surveys among the objectors, raising the level of debate from what might have been presented if there had been no need to counteract the environmental impact assessment.

The Minister's Decision

On March 25, 1982, Michael Heseltine, the Secretary of State for the Environment refused planning permission for the development of the Vale of Belvoir Coalfield. He did not accept the Inspector's view that there was no reason in principle why the mine buildings should not become acceptable visual elements in their proposed settings or that there would be no serious or widespread damaging effect on the area as a whole. Instead, he preferred the evidence given at the Inquiry that the Hose shaft towers would dominate a wide area and that the surface developments would be alien in the Vale. He said the proposals to develop a mine complex at Hose were unacceptable. He accepted the Inspector's conclusions that tipping operations at Hose and Saltby would in visual terms be totally unacceptable and that noise would be a problem. He was not convinced that the degree of need for coal demonstrated outweighed the adverse environmental effects.

He drew attention to the fact that the NCB had opted to stand or fall on a strategy of developing the whole coalfield as one project which led him to consider that to grant permission for only part of the development would be in effect granting a permission for a development significantly different in kind from the proposal which was the subject of the application. He did accept that there was a national need for more efficiently-mined cheap coal, and that regionally there was a strong case for the redeployment of miners from efficient but worked-out pits in North West Leicestershire, so the way was left open for the NCB to submit new planning applications relating to revised proposals.

In early 1983 the NCB submitted a planning application to Leicestershire County Council for planning permission for a coal mine at Asfordby which would employ 1100 miners and surface workers and 100 managerial and clerical staff. About 2.2 million tonnes of coal per year would be produced from coal reserves covering 60 square kilometres. Underground tunnels from Cotgrave Colliery would enable some of the seams in the north and west of the coalfield to be mined, and Nottinghamshire had already given planning permission for this development.

The Leicestershire County Planning and Recreation Sub-Committee approved the Asfordby Mine application on April 14, 1983, and the decision letter and Section 52 Agreement was dated May 17, 1983. Work started on the site in August, 1984. At 1984 prices the new mine will cost about £400 million, and it will take some nine years to construct so that coal production will not start until 1992 or 1993. The sinking of the two winding shafts for men and materials will take place during the first six years, while landscaping and the mine buildings will be completed by the seventh year. Production of coal is expected to commence by about the middle of the seventh year after the start of construction, with full production being achieved just over a year later. Anticipated productivity of 10.9 tonnes output per man shift will be four times the current national average.

Conclusions

The NCB must be congratulated on their pioneering attempt to introduce environmental impact assessment into the British planning system in such a complex, difficult and sensitive situation as in the Vale of Belvoir. On the face of it, the EIA appeared to provide a good case study of how environmental concerns influenced the location decisions for the proposed mines. In practice, however, reservations must be expressed about how much influence the EIA really had exerted upon the NCB as the developer. The choice of suitable mine sites would appear to have been strongly influenced by underground and technical mining considerations rather than surface conditions. In addition the NCB decision-makers, being engineers unfamiliar with physical planning, may have felt happier with the more quantitative measures of environmental impacts so that a fairly mechanical decision was reached.

With hindsight it is perhaps easy to criticize the NCB approach of 'all-or-nothing' with little prior consultations with the local planning authorities. The compromise or limited mining strategy which they eventually had to accept, namely an Asfordby mine plus limited expansion from the existing Cotgrave mine, could perhaps have been put forward in the first place if they had correctly predicted and accepted the environmental constraints of other potential mine sites. This would have retained options for future planning applications in five or ten years when the need for coal could have been predicted more easily. Had the economics of remote tipping of spoil been changed through the negotiation of special lower rates with British Railways or through a government subsidy, or if the London Brick Company could have been persuaded to buy spoil to fulfill its environmental obligations to restore the Bedfordshire brick-pits, the situation would have been different. Either of these modifications might have led to an earlier solution to the conflict between national and local issues, defusing the situation through undermining local opposition groups.

Yet it must be remembered that the NCB proposal almost succeeded. In contrast to Selby, the NCB was guaranteed a fight and could not ignore local pressure groups. In a rural area with many articulate middle-class commuters, and where few locals were likely to gain any benefits from the presence of a new coalmine in the short term, it was not surprising that the NCB felt obliged to show concern about environmental issues. Despite the criticisms fo the EIA, the arguments and site visits almost convinced the Inspector during the Public Inquiry, and perhaps if remote tipping or underground stowage of spoil had been economic, the NCB's planning application might have turned out to be acceptable.

What really upset the NCB 'strategy' however, was the change in government in 1979 prior to the start of the Public Inquiry. The results of the General Election could not have been predicted at the time of the planning application still less at the time when coal discoveries were first made in the Vale of Belvoir.

The arguments of the local pressure groups and politicians found greater favour with a Conservative Secretary of State for the Environment. Also national energy policies underwent a change, with the oil crisis appearing less severe and nuclear power stations finding more favour for electricity generation. National economic priorities had also changed with the monetarist Conservative government perhaps reluctant to fuel future inflation through a public sector investment programme. Ironically, the change in control of Nottinghamshie County Council from Conservative to Labour before the decision was made appeared to have had little effect since both administrations were concerned about the same issues. In the Leicestershire Country Council, Conservatives lost overall control but again the attitude of the new 'hung' council appeared to be essentially the same as at the time the application was originally submitted.

Many of the criticisms of the consultants' EIA stem from the absence of an agreed methodology for environmental impact assessment. Some observers claim many aspects of EIA are already incorporated into existing good planning procedures, especially in reports to local authority Planning Committees. Currently there is still not enough experience with EIAs to offer clear lessons about what effect the EEC Directive will have on the British planning system. Some doubts must remain about how effectively local planning officers can participate in EIAs without running into conflicts with elected members who are mindful of the fear of alienating large numbers of their local electorates.

The question of who should pay for the EIA remains unresolved. Local authorities, especially at the district level, are unlikely to have sufficient surplus manpower with the necessary expertise to be able to ignore financial constraints. Unlike many counties, Leicestershire had already submitted its Structure Plan, building on one of the earliest sub-regional plans in 1969, so it was able to undertake its own impact assessment. Even if an agreed EIA could have been jointly produced, a Public Inquiry would still have been held, though in a much-shortened form, thus some delays would still have been inevitable. The Minister would ultimately have taken the decision so it would still have entered the political arena. However, a clear lesson, from the Belvoir case is that the time taken to reach a final decision can only be reduced if a more comprehensive EIA is produced in the first place, which takes account of all the key issues of importance to developer and local planning authority and is properly integrated into decision-making procedures.

REFERENCES

Catlow, J., and Thirlwall, G. (1977) *Environmental Impact Analysis*. DoE Research Report No. 11. London: HMSO.

Dobry, G. (1975) *Review of the Development Control System: Final Report*. London HMSO.

Leicestershire County Council (1979) *North East Leicestershire Coalfield Impact Assessment*. Report of the County Planning Officer, February, 1979.

Leonard and Partners and Thyssen (G.B.) Ltd. (1976) *Vale of Belvoir Project Feasibility Report*. Report by the Consultants to the N.C.B. (South Notts. Area), especially Vol. 2, Surface Works Report and Plan.

Leonard and Partners and Thyssen (G.B.) Ltd. (1977) *Belvoir Prospect*. Report by the Consultants to the N.C.B. (South Notts. and South Midlands Areas), especially Vol. 2: Surface Works Report, Appendices, and Plans.

Mann, Michael, (1981) *The Vale of Belvoir Coalfield Inquiry* (The North East Leics. Prospect). Report: Applications for Planning Permission by the N.C.B. Presented to the Secretary of State for the Environment on 26 November 1980. London: HMSO.

Thirwall, G. (1978) EIA – taking stock. *Built Environment*, 4 (2), pp. 87–93.

Williams, K. G. (1978) East Midlands record: proposed exploitation of the North-East Leicestershire (Vale of Belvoir) Coalfield. *East Midland Geographer*, 7 (2), pp. 83–88.

Williams, K., Hills, P., and Cope, D. (1978) EIA and the Vale of Belvoir coalfield. *Built Environment*, 4 (2), pp. 142–51.

Chapter Four

The Environmental Impact Assessment Directive of the European Communities

R.H. WILLIAMS

In March 1985, after many years of deliberation, the Environment Council of the European Communities agreed to adopt a 'Directive concerning the assessment of the environmental effects of certain public and private projects'. This Directive, commonly referred to also as the Environmental Impact Assessment (EIA) Directive, is an integral part of the European Communities (EC) environment policy, and represents a major step forward in implementing this policy. It is also the first example of Community land-use planning legislation to be incorporated into national physical planning systems.

Other chapters have demonstrated how, and in what ways, EIA is now playing a role in the planning process. This chapter shows how this role may be developed so that the planning process can make a greater contribution to the prevention of environmental pollution, not only in the United Kingdom but throughout the EC, as a result of the adoption of the Directive. In doing so, it will consider the rationale for the promotion of an environment policy by the EC, the place of EIA within it and the benefits it is hoped will accrue from its introduction at a supra-national level. The chapter then outlines the provision of the Directive itself, the process of incorporation into existing national planning systems and the potential significance of the Directive for the planning process.

Why an EC Environment Policy?

The European Communities (Coal and Steel, Euratom and the European Economic Community) owe their origin to a desire for political and economic integration, and no reference is made to an environment policy in the Treaty of Rome. However, a Community environment policy has been in operation since

1973. It is based on the premises that the Community needs to be concerned with the quality of life of its population, as well as economic integration and growth; that pollution tends by its very nature to cross national frontiers; and that there can be direct economic benefits from environmental protection, especially in the agriculture, forestry, fisheries and tourism sectors and through the stimulation of clean technologies and manufacture of the necessary equipment (Commission of the European Communities, 1979). The legal basis for an environment policy rests on Articles 100 and 235 of the Treaty of Rome, concerned with harmonization, and initiation of policies not otherwise specified in the Treaty (Williams, 1984, chapter 12). Three Action Programmes on the Environment have been approved since 1973, embodying the elements of the Community environment policy. These have covered the periods 1973 - 76, 1977–81 and 1982 - 86. Another programme was begun in 1987, a year which was designated the Year of the Environment. An initial indication of their thinking has already been issued by the Commission (Commission of the European Communities, 1986).

The Action Programmes have included the measures which the Commission has proposed to bring forward to the Council of Ministers in each period, based on analyses of the problems being faced in the Community. Inevitably, however, negotiation of some measures has been protracted. The proposed EIA Directive, for example, figured in the second Action Programme and was carried forward into the third Action Programme before being adopted during the latter period.

Principles underlying the policy

Environment policy measures are introduced on the basis of two fundamental principles: that the polluter should pay; and that prevention is better than cure. Implementation is guided by the additional administrative principle that action should be taken at the appropriate geographical level of jurisdiction: local, regional, national, Community or international (Fairclough, 1983). These are considered in turn, concentrating on the second principle because this provides the basic rationale for the link between the EC environment policy and the land-use planning process.

A conflict of interest is likely to exist between the potential polluter, who may want to maximize profits, and regards costs of anti-pollution measures as being externalities to be avoided if possible, and the interests of the community or district suffering pollution. The principle that the polluter should pay for the costs of emission control or other anti-pollution measures, and not regard these as externalities outside his responsibility, has therefore been a feature of the environment policy from the Community's first Action Programme on the Environment, in the interests of those potentially affected by pollution (Commission of the European Communities, 1979; Fairclough, 1983).

The principle that prevention is better than cure was put forward because of the high cost of clearing up after major pollution. Dramatic evidence of this occurred within the EC after the Amoco Cadiz oil spillage and the Flixborough and Seveso disasters. The last also created the political impetus for the adoption of a Directive on the storage and use of dangerous substances in industrial plants, known as the Seveso Directive (Commission of the European Communities, 1982a). The Chernobyl disaster in April 1986 put the environment firmly back on the political agenda. Although outside the jurisdiction of the EC, it demonstrated the international nature of pollution and will have an impact on EC public opinion and policy.

The second Action Programme explicitly adopted the principle that 'prevention is better than cure' as a fundamental basis for environment policy, on the basis that the policy should prevent 'the creation of pollution of nuisances at source, rather than trying to counteract their effects' (Commission of the European Communities, 1977, p. 50). As Wood has demonstrated, the use of land-use planning procedures is a major way in which this principle can be put into practice. (Wood, 1979). The Action Programme itself saw this link as a means of strengthening the environmental dimension of planning, setting an objective 'to ensure that more account is taken of environmental aspects in town planning and land use' (Commission of the European Communities, 1977).

Several proposals contained in the second Action Programme and also in the third Action Programme are of a preventive type. In the latter they are related to the practice of land-use planning more explicitly than ever before. As the third Action Programme puts it:

> Land in the Community is a very limited and much sought after natural resource. The way in which it is used very largely conditions the quality of the environment. Physical planning is therefore one of the areas where a preventive environment policy is very necessary and very beneficial. (Commission of the European Communities, 1983, Art 26).

A very wide range of interests may be affected by physical planning measures of this sort since they are designed to control hypothetical future pollution. Consequently, the number of interest groups seeking to resist or dilute a preventative planning procedure, especially on a European scale, is considerable, and in certain cases includes national government departments as well as non-government public and private bodies. In addition, there is by no means general public or political support for extending, or appearing to extend, the scope of town planning and environmental protection procedures, and proposed controls may be interpreted as placing further hurdles in the way of developers who are offering much needed jobs and economic stumulus. This was clearly reflected in the Danish veto on the Directive during 1983–1985 (see below), and is no doubt a factor in the United Kingdom government's luke-warm attitude.

In addition, all member states have already got a system of town and country planning (Williams, 1984). Although these have broadly similar overall objectives, the differences between them both in procedures and in effective policies are very great, for a variety of legal, constitutional and historical reasons. It is not surprising, therefore, that the formulation and adoption of an acceptable land-use planning measure took a long time.

The Commission, in its progress report (Commission of the European Communities, 1980a) acknowledged as much. The technical complexity of many of the measures proposed, and limited staffing resources are identified as reasons for slow progress. Other reasons put forward include the existing variety of institutional and administrative procedures and responsibilities in the different member states, and the uncertainty of political will. The progress report refers to the need for political impetus if progress towards better environmental protection is to be achieved, referring to the fact that difficulties may sometimes arise if the 'political will is uncertain' and to 'the half-hearted political response to oil pollution at sea' (Commission of the European Communities, 1980a p. 5). It is clear from the report that general preventive measures affecting potentially a wide range of interests, such as land-use planning measures, were among those where sufficient political support for progress was lacking at that time.

The lack of progress in adopting land-use planning measures to control pollution during the course of the second Action Programme on the Environment has meant that many of the proposals envisaged then also found a place in the third Action Programme (Commission of the European Communities, 1983). However, certain changes of emphasis are apparent in the third Programme. It placed greater emphasis than earlier programmes on jobs, on a trans-sectoral approach linked to the EC's regional and industrial policies, co-ordination of specific anti-pollution measures, monitoring, co-operation with other national or international agencies, and with research, education and information exchange. It continued to emphasize prevention, but in doing so made much more frequent and explicit reference to the role of physical land-use planning. The argument that environmental protection is not a luxury to be afforded only in times of prosperity, but a necessary part of economic development at all times, was put forward more strongly than in earlier programmes. The Commission has already indicated that the link with the economy will be increasingly strongly emphasized in the future (Commission of the European Communities, 1986).

The administrative guidelines that action should be taken by authorities having jurisdiction at the appropriate geographical scale has a sound practical base in view of the limited staff resources of the Commission. In the case of the EIA proposal, this rule makes sound operational sense, facilitating the incorporation of the EIA process as required by the Directive into the work of local planning authorities.

Harmonization

In the case of the EIA Directive, as with many other EC proposals, the need to adjust existing national systems in the interests of harmonization at a European scale has not always been fully understood. At the time when consultations were being carried out on the proposed Directive, many UK organizations commented critically on its implications for existing British practice, fearing imposition of rigid environmental quality standards. In adopting this stance, there was a widespread failure to recognize that harmonization is not only in the overall EC interest, but also in the national interest of any member state seeking to adopt high standards of environmental quality. Without harmonization there is a danger of distortion of competition for industrial development which may affect the national economy or environment, for reasons which are argued below.

The willingness of the potential polluter to bear any additional costs imposed as a result of the requirement to adopt pollution control measures depends on where the economic interests of the developer lie. In the case of a locationally mobile or multi-national company, there is likely to be the temptation to propose an alternative location for a development where pollution control requirements may be less stringent and less expensive. This temptation is likely to be particularly strong if the alternative site offers equally good access to the same market. In Europe, locationally mobile multi-site or multi-national firms can seek to avoid costs of anti-pollution measures by choosing a location in another member state of the EC where the approval procedures are more easily satisfied and lower standards of environmental protection allowed to prevail. The developer nevertheless still has access to the same market, being within the common external tariff. In this way, pollution havens may emerge, and distortions in competition occur as firms have to comply with very different environmental standards, while competing in the same market (Williams, 1983). It is therefore in the interests of the Community and of member states with high environmental standards, such as Federal Germany, to harmonize authorization and control procedures throughout the EC, in order to avoid economic disadvantage as a result of high national standards.

The common environmental assessment procedure now to be introduced is intended to have the effect of reducing the disparities in environmental standards between member states, by bringing all national authorization procedures up to an acceptable minimum standard, in order to counteract any temptation to locate in a pollution haven.

The European scale of operation of the proposed procedures will greatly enlarge the geographical scale at which spatial variations in planning and pollution control might affect the distribution of industrial or other polluting activities. Blowers (1980) refers to the power of the environmental protection movement to influence the distribution of industry and the creation of a new geography of industry, with pollution havens where pollution controls are being applied less strictly, for either political or procedural reasons, and to which polluting activities therefore tend to locate. Harmonization of approval procedures on a European scale would have the effect of reducing procedural variations as a factor in location decisions within the EC, but would not necessarily have any effect on variations due to differences in political attitude and political will to enforce strict pollution standards.

Place of Environmental Assessment

The concept of environmental assessment of projects fits very well into the logic underlying the Community environment policy. In essence, it is a procedure whereby the environmental consequences of a proposed development are assessed at the time when authorization of the development is sought. The prospective developer is required to submit sufficient details of the proposal to allow the environmental consequences, and the adequacy of any remedial measures to be assessed prior to authorization of the development. The authority competent to grant permission is required to assess the project, grant permission to develop only if it is satisfied that any adverse environmental consequences are alleviated as far as possible, and impose any necessary conditions to ensure that any environmental damage from the project is minimized, and the danger of subsequent pollution removed as far as possible. It is thus a procedure which embodies the principle that prevention is better than cure, and imposes on the developer the discipline of considering the environmental consequences of a proposed development at the time when it is in the developer's interest to pay for any necessary remedial measures in order to obtain authorization for the development.

Environmental assessment need not necessarily be seen as an integral part of land-use planning and control of development. It could be associated with separate environmental and public safety procedures, for instance. However, the normal arrangement will be for environmental assessment to take place alongside the existing procedures for authorization of land use and development, in the case of the United Kingdom as part of the development control responsibilities. It will, in fact, be the first example of the use of a physical planning procedure operating throughout the Community as a means of pollution control.

Although the requirement to undertake environmental assessment is embodied in an EEC Directive, the actual operation of the procedure is the responsibility of the authorities designated for this purpose by member-states. Thus it will come into operation in accordance with the administrative guidelines referred to above. In the United Kingdom it will be operated by local planning authorities, and a similar level of government is likely to be responsible elsewhere in the Community.

Environmental assessment of individual projects had been intended by the

European Commission to be complemented by similar procedure to assess plans and programmes. Proposals to implement these other two forms of environmental assessment have not yet been tabled, although the European Parliament did resolve that consideration of plans should be incorporated in the present proposal (Commission of the European Communities, 1981). This amendment was not made by the Commission, however. Assessment of plans might operate in quite a straightforward manner in the case of local building plans such as the German *Bebauungsplan* or the Dutch *Bestemmingsplan*, and other similar plans in other countries, but it would be a much more complex matter in relation to British development plans, with less certain benefits in view of their indicative nature and discretionary relationship with development control.

Assessment of programmes might have a more tangible, but politically sensitive role, in the British case. Concern has mounted during recent years over the lack of opportunity for public debate into the merits of national programmes of development for projects such as motorways, nuclear power stations, new coal fields or the third London airport. It has often been argued that there is a need for a public debate or inquiry into the merits and justification of major public sector programmes such as those for motorways or power stations at the time when they are being considered in principle, and before site specific proposals are brought forward for consideration. Normal practice is for the debate at this earlier stage to take place only in Parliament, if at all, and for public inquiries to be held only into site-specific proposals. Such public inquiries can easily prove unsatisfactory and difficult to manage when objectors are really concerned to challenge the principle of a national programme, rather than the local element of that programme.

Debate and detailed examination of any major set of proposals at the non site-specific programme stage is something that has happened in Britain in the case of the Roskill Commission of Inquiry into the siting of the third London Airport (Roskill, 1971) and issues of national policy were considered at the Sizewell Inquiry held in 1984. Statutory provision for this type of exercise, by means of a Planning Inquiry Commission, was incorporated in the 1968 Town and Country Planning Act, but no such Commission has been set up subsequently.

An extension of the principle of environmental assessment from the present Directive, which relates to projects and is therefore tied to site-specific proposals and applications for planning permission, to programmes would go a long way towards filling the gap described above by enabling a series of projects to be assumed in principle, before individual site-specific proposals are put forward for local consideration. The Fourth Environmental Action Programme, published in 1986, though not formally adopted at the time of writing, contains a general intention to develop further the existing EIA proposals. Some consideration is also being given to the relationship between European environmental policy and the European Regional Development Fund (ERDF).

The EIA Directive

In June 1980 the Commission tabled a proposed directive 'concerning the assessment of the environmental effects of certain private and public projects'. (Commission of the European Communities, 1980b). The version subject to this formal proposal was itself the product of a long series of studies and consultations over several years, generating numerous revisions of the draft.

This document consisted of a substantial explanatory memorandum, plus the Articles of the proposed text, and three Annexes. The Articles and the Annexes are the parts which become Community legislation upon adoption.

The legal basis of the proposal is Article 100 of the EEC Treaty, under which the Council may adopt directives for the approximation of laws directly affecting the functioning of the Common Market.

The 1980 proposal was presented to the Council of Environmental Ministers, who were required to consult national governments, the European Parliament and the European Economic and Social Committee before they could take a decision on it. This process took place during 1980–82, as a result of which a revised proposal was prepared and submitted to the Environment Council in March, 1982 (Commission of the European Communities, 1982a) after widespread consultations.

The evidence presented to the House of Lords Select Committee of the European Communities is the principle source available indicating considered responses and attitudes to the directives (House of Lords, 1981). Many organizations and interest groups took advantage of the opportunity to make representations to the Select Committee, or give oral evidence, and their final report is a very thorough analysis of the proposal, forming very favourable conclusions for those advocating adoption of the Directive. The attitude of the government in the subsequent Commons debate was markedly less enthusiastic.

A consistent line of argument from both government and industrialists against the provisions of the Directive concerns the question of standards and formalized procedures. There seemed to be an underlying fear that the Directive is the product of an alien system, and would cause the imposition of rigid procedures and fixed standards of environmental quality. These fears were expressed by several witnesses to the House of Lords Committee, including the Department of the Environment, the Association of Metropolitan Authorities and the Confederation of British Industry, all of whom indicated that they would prefer an informal code of practice rather than a directive (House of Lords, 1981). All accepted the desirability of environmental protection and pollution control but argued in effect that this was already achieved by existing legislation and the development control system.

The British government was sympathetic to this point of view, and was one of the less enthusiastic national governments in respect of this proposal. However, a draft acceptable to the British government was negotiated by the end of 1983. Final approval of the Directive by the Council of Ministers did not take place until the March 1985 Environment Council, due to objections from Denmark. A fuller discussion of the proposals and the process of adoption may be found in Williams (1983 and 1986).

The Directive as Adopted

The Directive as approved by the Environment Council in March 1985 was formally issued on June 27, 1985, as Directive 85/337/EEC, *On the assessment of the effects of certain public and private projects on the environment* published in the *Official Journal*, No. 175, of July 5, 1985.

The text of the Directive is not long. It consists of fourteen Articles, some being purely procedural, and three Annexes. It follows broadly the pattern of the earlier proposals referred to above (Commission of the European Communities, 1980, 1982a) but with a number of highly significant variations (see also Williams, 1986).

Article 1 defines the terms and scope of the Directive. A very wide range of projects could be included, and the term developer is defined so as to include both private developers and public authorities responsible for initiating projects. Development consent is defined as authorization to proceed with the development of the project by the competent authority designated by national government to take this responsibility. In effect, this allocates responsibility to the local planning authority from whom planning permission is sought. The position is more complex when projects which do not require approval from the local planning authority are considered, although these are in principle equally subject to assessment.

In a significant amendment to the Commission's earlier proposals, Article 1 now excludes from assessment national defence projects and projects authorized by a specific Act of Parliament. The new London Docklands Light Railway was authorized by this means and would therefore have not been subject to the Directive, and it is intended that works associated with the Channel Tunnel will also be authorized by Act of Parliament.

The essence of the directive is contained in Article 2(1):

> Member States shall adopt all measures necessary to ensure that, before consent is given, projects likely to have significant effects on the environment by virtue inter alia, of their nature size or location are made subject to an assessment with regard to their effects.

The text goes on to indicate the procedure for exemption (Art 2[3]), scope of the assessment (Art 3), the information to be supplied by the developer (Art 5), public participation (Arts 6 and 9), consultation with other member states (Art 7) and supply of information to the Commission, who are to monitor the operation of the directive and report on it to the European Parliament (Art 11). The Directive must come into operation within three years of notification, i.e. by July 1988 (Art 12), not within two years as originally proposed.

The potential scope of an assessment is very wide: Article 3 refers to:

the direct and indirect effects of a project on the following factors:
- human beings, fauna and flora,
- soil, water, air, climate and the landscape,
- the inter-action between the factors mentioned in the first and second indents,
- material assets and the cultural heritage.

A detailed list of the type of information likely to be required is provided in Annex III. This includes descriptions of the site, project and planning context, technical data on the impact on the natural environment, built environment and the population, proposals to minimize these impacts, and a non-technical summary.

Article 4 and the two Annexes referred to in it define the type of projects to be subject to assessment. Projects in Annex I are to be subject to the full procedure, unless exemption is granted. Exemption is intended to apply only very occasionally, and must be justified to the Commission, other members states and the public.

Annex I lists crude oil refineries, thermal and nuclear power stations, radioactive waste disposal, integrated steel works, asbestos extraction and processing, integrated chemical installations, major roads, railways and airports, port development including major inland waterways and waste disposal facilities for toxic and dangerous materials.

The type of project listed in Annex 1 is in general the type of project which would be subject to very detailed scrutiny by planning authorities in any event. Consequently, much if not all of the information required under this Directive would be required by the local authority at some stage anyway, and it may therefore be argued that this proposal corresponds to existing good practice, and will have the effect of levelling all authorities up to this standard.

The Commission originally proposed that the second list of projects (Annex II) would be subject to a simplified form of assessment. Although the proposal was not very specific about the nature of a simplified assessment, it was clear that some form of assessment would have been required. However, the final text refers in Article 4(2) to the assessment of projects in Annex II 'Where Member States consider their characteristics so require'. Thus, discretion rests with the national authorities. Annex II is a much longer list, with twelve categories of project, including agriculture, extractive industry, energy, a variety of manufacturing processes, and infrastructure projects including transport, industrial estates, urban development and mains services. If all projects in Annex II were to be subject to assessment, even in simplified form, the environmental assessment process would operate much more extensively throughout the operations of the planning system than would be required by Annex I alone. It would also bring within a form of planning control certain agricultural processes not currently subject to the UK Town and Country Planning Acts. However, the discretion allowed to member states referred to above will no doubt limit the impact of the Directive on Annex II projects, but to what extent is not clear at the time of writing.

The Operation of the Directive

As a Directive, it is binding as to the means to be achieved, but member states are themselves responsible for giving it legislative effects. Therefore, within the three-year period from notification in July 1985, member states are required to enact any necessary legislative measures to introduce the system and integrate it into existing procedures.

In the United Kingdom, no measures have yet been proposed to Parliament, but procedures for implementing the Directive are being considered by a working party established by the Department of the Environment. An advisory booklet will be issued by the Department, and the Directive will be incorporated into UK law by means of an Order under the European Communities Act 1972 and if necessary by supplementary Orders under the Town and Country Planning or other Acts.

It is too early, therefore, to comment on how the directive will be incorporated into planning practice in the United Kingdom, although some general observations may be made. The Directive itself contains a mixture of articles which are very wide ranging in their scope and articles allowing considerable freedom for member states to operate the assessment procedure in a way which suits them and is compatible with existing procedures. This represents a compromise between the objectives of the Commission, seeking to harmonize and strengthen procedures throughout the EC as part of its environment policy, and the wishes of member states to retain freedom of action.

Article 5, for instance stresses that the information required must be relevant and reasonable, so that the assessment does not become overwhelmed by detail, whereas Article 3 lists a very wide range of factors to be taken into account.

Article 6 requires extensive consultation with other appropriate agencies, and public participation. The form taken by the latter is to be determined by the member states, but it is intended that the assessment made by the prospective developer is in the form of a document available for public scrutiny.

Early experience of implementing legislation requiring Enviromental Impact Statements in the United States gave rise to fears that implementing this proposal would cause delays and increase developer's costs. It is possible, however, that the discipline of preparing this assessment in advance of submitting a planning application could speed up the approval process and save money for the developer. In the case of major controversial proposals under present United Kingdom legislation extra information may well be requested by local planning authorities or the inspector during a public inquiry, thus causing delay. If an assessment is required as part of the normal application procedure, all necessary information should be assembled while the application is being prepared and subsequent delays could therefore be avoided.

The best documented support for this point of view comes from the British Gas Corporation, who claimed that they have saved a total of £30 million over ten years as a result of preparing full assessments of their projects prior to seeking authorization, thus enabling the application to be considered more quickly (House of Lords, 1981, p. 52).

In one respect, consultation could be extended more widely than is normally the case at present. Article 7 requires, where significant cross-frontier environmental effects may be attributable to a project, that the other appropriate member states be supplied with the assessment and consulted. This will normally apply along land frontiers, but it is not necessarily irrelevant for the United Kingdom. For instance development near the English Channel may create development pressures on the opposite coast. More contentiously, if authorities in Denmark or Germany could establish a link between acid rainfall there and specific emissions in the United Kingdom, projects likely to cause such emissions could be subject to this procedure. Thus, an international dimension may be added to the work of some local planning authorities. Also, of course, any nuclear installation is likely to be regarded as being of trans-frontier significance after Chernobyl. It is noticeable already how often such installations are located near frontiers.

There is concern in several countries about the environmental and ecological consequences of modern farming techniques, the effects of hedgerow removal and land consolidation, and the impact on the countryside of various installations for intensive stock rearing. If all forms of agricultural project listed in section 1 of Annex II were to be subject to a form of environmental assessment, a significant extension of planning control into this sector would take place, at least as far as the United Kingdom is concerned. However, the discretion allowed over Annex II projects in Article 4(2) will no doubt be exercised so as to limit any such extension of planning.

Conclusion

Environmental impact assessment has not been the subject of specific legislation in the United Kingdom, although it has in certain other member states including France, Netherlands and Ireland (Haigh, 1983; Wood, chapter 10). However, there is extensive experience of operating environmental impact assessment in the United Kingdom within the framework of the Town and Country Planning Acts, as other chapters have demonstrated. The Directive itself has been eagerly awaited by the environmentalist lobby, who have had high hopes that it will serve to strengthen the pressures which can be exerted to resist environmentally damaging development.

However the proposed Directive is by no means an environmentalist's charter, or wholly against the interests of developers, in spite of the general welcome from the environmental lobby and suspicion from developers and government interests.

Within the United Kingdom, it will reinforce existing good practice and no doubt have the effect of encouraging planning authorities to follow the example of those which already have extensive experience of environmental assessment.

At the EC level, it will be a step in the direction of strengthening the planning control systems in those member states, especially in Southern Europe, where they are relatively weak. It could have a modest, but discernable effect in strengthening the hand of interests wishing to control pollution. It would extend the scope of planning control, and would create more opportunities for environmental interests to identify potential sources of pollution and for more effective pursuit of higher environmental standards by the competent authorities. It also provides the means whereby some harmonization of standards and reduction of pollution havens could be achieved.

Much will depend, of course, on the extent to which common standards are adopted at national or community level. At present, it appears quite possible for a local authority determining planning applications requiring assessment to make exactly the same decision as they would have done without the Directive. However, the impact of its adoption would be much greater, and more equality of environmental standards in different authorities would be achieved if common standards were to be adopted.

Its real significance lies in the fact that it is the first major attempt to establish a common procedure throughout the Community to prevent environmental damage, through the medium of land-use planning and development control procedures. It may well prove to have only a modest impact on the type of development permitted, and the conditions attached to planning consent, in mature planning systems such as that of the United Kingdom, but to be of greater importance as a means of boosting weaker planning systems.

REFERENCES

Blowers, A. (1980) *The Limits of Power*. Oxford: Pergamon.

Commission of the European Communities (1973) *Official Journal* C112, 20 December, pp. 1–53.

Commission of the European Communities (1977) *Official Journal* C139 , 13 June, pp. 1–46.

Commission of the European Communities (1979) *State of the Environment. Second Report*. Brussels: CEC.

Commission of the European Communities (1980*a*) Progress made in connection with the environment action programme and assessment of the work done to implement it. Com (80) 222 Final.

Commission of the European Communities (1980*b*) Draft directive concerning the assessment of the environment effects of certain public and private projects. Com (80) 313 Final.

Commission of the European Communities (1981) European Parliament Document 1–569/81, 21 October.

Commission of the European Communities (1982*a*) Proposal to amend the proposal for a Council Directive concerning the environmental effects of certain public and private projects. Com (82) 158 Final.

Commission of the European Communities (1982*b*) *Official Journal* L230, 5 August, pp. 1–56.

Commission of the European Communities (1983) *Official Journal* C46, 17 February, pp. 1–22.

Commission of the European Communities (1985) *Official Journal* L175, 5 July, pp. 1–48.

Commission of the European Communities (1986) New directions in environment policy. Com (86) 76 Final.

Commission of Inquiry into the Third London Airport (1971) *Report* (The Roskill Report). London: HMSO.

Fairclough, A.J. (1983) The community's environmental policy, in Macrory, R. (ed) *Britain, Europe and the Environment*. London: Imperial College Centre for Environmental Technology, pp. 19–34.

Haigh, N. (1983) The EEC Directive on the environmental assessment of development projects. *Journal of Environment and Planning Law*, September, pp. 585–95.

House of Lords (1981) *Environmental Assessment of Projects*. Select Committee on the European Communities, 11th Report 1980–1, Paper 69. London: HMSO.

Williams, R.H. (1983) Land use planning, pollution control and environmental assessment in the EC environment policy. *Planning Outlook* 26 (2) pp. 54–59.

Williams, R.H. (ed.) (1984) *Planning in Europe*. London: George Allen & Unwin.

Williams, R.H. (1986) EC environment policy, land use planning and pollution control. *Policy and Politics* 14 (1), pp. 93–106.

Wood, C.M. (1979) Land use planning and pollution control, in O'Riordan, T. and D'Arge, R.C. (eds) *Progress in Resource Management and Environmental Planning*. Chichester: Wiley.

Chapter Five

The Genesis and Implementation of Environmental Impact Assessment in Europe

CHRISTOPHER WOOD

Environmental impact assessment refers to the assessment of the environmental impacts likely to arise from a major action (i.e legislation, a policy, a plan or a project) significantly affecting the environment. The formalization of EIA stems from the US National Environmental Policy Act 1969 (NEPA) and subsequent legal rulings and practice which require developers to demonstrate that they have carried out an assessment by publishing an environmental impact statement (EIS) describing in detail the environmental impacts (including use of resources) likely to arise from the development.[1]

It must be emphasized that EIA is intended to form an integral part of the process of formulating and considering an action (for example, when building a new power station) which may then be modified or even abandoned to mitigate the forecast environmental impacts (for example, by choosing another location, a different fuel or another electricity transmission system). EIA should therefore be seen as an environmental management tool within the planning process: EIA is not, and can never be, a replacement for land-use planning.

EIA has been used by many international agencies (e.g. the World Bank, the United Nations Environment Programme) and by many countries (e.g. Canada, Australia, Japan). Its writ is spreading. This chapter is concerned with its adoption by the Commission of the European Communities, with EIA in the European member states and with attitudes to EIA in the United Kingdom. Now that the Commission has approved a directive on EIA, some form of assessment will become mandatory in the United Kingdom. The final part of the chapter deals with some of the implications of a mandatory system for United Kingdom planning practice.

The European Directive

The European Communities have espoused the principle that prevention is better than cure in environmental protection: 'the best environmental policy consists in preventing the creation of pollution or nuisances at source rather than subsequently trying to counteract their effects' (CEC, 1977). The Commission has stated that 'too much economic activity has taken place in the wrong place, using environmentally unsuitable technologies' (CEC, 1979, p. 49) and that 'effects on the environment should be taken into account at the earliest possible stage in all the technical planning and decision-making processes' (CEC, 1977). Because of this concern to anticipate environmental problems, and hence to prevent or mitigate them, the Commission became interested in environmental impact assessment in the early 1970s.

Following a number of EIA research programmes, the Commission decided that an EIA system should meet two sets of objectives:

1. To ensure that distortion of competition and misallocation of resources within the European Economic Community (EEC) are avoided by harmonizing controls.

2. To ensure that a common environmental policy is applied throughout the EEC.

The Commission issued its first preliminary draft Directive in 1978. After twenty such drafts, not all of which were released, and substantial consultation (this is reliably reported to have been the most discussed European draft directive ever), the Commission put forward a draft to the Council of Ministers in June 1980 (CEC, 1980). Before the finally approved version this was the only version of the Directive to have been published in the *Official Journal of the European Communities*.[2]

The draft Directive specified that projects likely to have a significant effect on the environment were to be subject to an EIA. Such an assessment was obligatory for nearly all projects, other than modifications to existing installations, in certain specified categories which were listed in Annex 1. There were some thirty-five of these types of projects, grouped under the headings: extractive industry; energy industry; production and preliminary processing of metals; manufacture of non-metallic mineral products; chemical industry; metal manufacture; food industry; processing of rubber; and building and civil engineering.

An EIA was also obligatory for projects in certain other specified categories listed in Annex 2 of the Directive and for substantial modifications to Annex 1 projects, but subject to criteria and thresholds to be established by member states. Annex 2 included agricultural and forestry practices as well as many industries not encompassed by Annex 1. In addition, an EIA was to be required for any other projects outside the above categories where a significant environmental impact was likely to occur. There were provisions for Commission co-ordination of criteria and thresholds and for a simplified form of assessment in certain cases. These proposals left considerable discretion with the member states in deciding the precise coverage of the EIA system to be adopted.

The developer was to bear the primary responsibility for supplying all the relevant basic information required in an environmental impact study. At the same time, it was envisaged that the 'competent authority' would often need to assist the developer in the preparation of the study. The authority also had the responsibility of checking the information supplied which was to include:

(a) a description of the proposed project and reasonable alternatives to it;
(b) a description of the environment likely to be significantly affected by the project;
(c) an assessment of the project's likely significant effects on the environment;
(d) a description of any environmentally mitigating measures that are proposed;
(e) an indication of the likely compliance with existing environmental and land-use plans and standards for the area;
(f) a justification of the rejection of reasonable alternatives to the proposed project where these are expected to have less significant adverse effects on the environment;
(g) a non-technical summary.

Annex 3 specified the required content of the assessment in more detail. The kinds of impact to be considered included those arising from the physical presence of the project, the resources it used, the wastes it created and its likely accident record. Economic and social effects were excluded (although social impacts could be incorporated in the EIA).

There were fairly full provisions for consultation. The competent authority had to publish the fact that the application had been made, make all the environmental documentation available to members of the public and make arrangements for concerned parties to present their views. The competent authority had then to make its own final assessment and publish it (unless permission was refused on other than environmental grounds). This publication was to contain the assessment itself, a summary of the main comments received, the reasons for granting or refusing planning permission and the conditions, if any, to be attached to the granting of the permission. In the event of the project being authorized, the competent authority was expected to check periodically whether any conditions attached to the approval were being satisfied, and whether the project was having any unexpected environmental effects that might necessitate further measures to protect the environment. It was intended that assessment should eventually be extended from projects to plans and programmes. The system outlined bears a strong resemblance to most comprehensive environmental impact assessment systems.[3]

The Council of Ministers did not approve the draft Directive in June 1980. The British government, for reasons elaborated below, was reluctant to accept the imposition of a mandatory system of EIA, at least in the form set out in the published Directive. During subsequent negotiations, between the Commission and the British government several of the more controversial aspects of the Directive were deleted to meet the British position. In the course of the amendments, many types of industry were shifted from Annex 1, where their assessment would be compulsory, to Annex 2, where their assessment would be much more discretionary.

The approved version of the Directive (CEC, 1985) has limited the Annex 1 projects to oil refineries, large coal gasification and liquefaction plants, large power stations, radioactive waste disposal sites, integrated steel works, asbestos plants, integrated chemical plants, motorways, railways and large airports, ports, canals and toxic waste disposal facilities. The list of Annex 2 projects has grown substantially but the requirement for Commission co-ordination of criteria and thresholds has been dropped.

The main text of the Directive no longer mentions the need to discuss alternatives, deletes the provision involving compliance with land-use plans, standards, etc., but demands the identification, description and assessment of the direct and indirect effects of the project on human beings, flora and fauna, on soil, water, air, climate and landscape, on the interactions between these, and on material assets and the cultural heritage. The requirements for early provision of information to the public and the publication of the competent authority's own assessment have also been substantially weakened and that relating to monitoring has been deleted. Annex 3, specifying the desirable (rather than the required) content of the information supplied by the developer, reflects the reduced scope of the requirements. It does, however, expand the range of impacts to include those on 'population', thus presumably widening the scope of the assessment to socio-economic considerations rather than just the effects of the project on the physical environment.

The net effect of these changes has been the emasculation of the provisions in the early drafts of the Directive. These early versions were themselves criticized because they were over cautious, and did not contain provisions for the Commission to monitor and oversee the EIA system effectively, let alone use it to make substantive and constructive inputs to problem solving (Wandesford-Smith, 1979). If such observations had validity then, how much more must they apply now? There are however, some grounds for believing that the effects of the Directive may be more positive than these criticisms imply.

The British government withdrew its objections to the amended draft Directive at the Council of Ministers' meeting in late 1983 but the Danish Minister was unable to accept the Directive because a review of Danish

Table 1. EIA arrangements in the member states of the EC (mid-1980s).

BELGIUM: To date, one regional EIA provision has been enacted and a number of draft proposals for EIA legislation has been prepared at the national and regional levels of government. However, certain elements of an EIA system already exist in a number of environmental regulations relating to the granting of permits, licences and authorizations. The most important of these regulations are the General Regulation for Protection of Labour (approved by decree in 1946), which controls 'dangerous, dirty and noxious' establishments, and the Organic Law on Land Development and Town Planning (1962) which provides for different kinds of land-use and building plan, and for the granting of building permits.

DENMARK: There are no specific legislative provisions relating to EIA at the present. However, some of the elements of an EIA system are already in existence, notably within the pollution certificate system operated under the provisions of the Environmental Protection Act (1973) and in the preparation of regional plans and of supplements to these, under the provisions of the National and Regional Planning Act (1969).

FEDERAL REPUBLIC OF GERMANY: There are no specific legislative provisions relating to EIA at the present. However some kind of environmental evaluation is required in German environmental law in relation to the licensing of certain industrial plants, water uses and nuclear power stations and in the plan approval procedures for a wide range of infrastructure projects. By Federal Cabinet decision (1975) there is also a general obligation to carry out 'examinations for environmental compatibility' in the case of public measures undertaken on behalf of the federal government. At state level, a small number of länder (Saarland, Berlin, Bavaria) has adopted EIA procedures, some (Hamburg, Hesse) are in the process of doing so, and others (Schleswig-Holstein and North Rhine-Westphalia) await approval of the Community's directive.

FRANCE: Of all the Member States in the European Community, France currently possesses the most extensive, formal EIA system. Law 76–629 (1976) (implemented by Decree 77–1141 (1977)) makes provisions for impact assessments preceding decisions on project proposals. Provisions for an EIA are also included in Law 76–663 (1976) (implemented by Decree 77–1133 (1977)) concerning installations registered for purposes of environmental protection. It has been estimated that as many as 8,000 environmental assessments are prepared each year and these EIA reports are considered to be complementary to the technical, economic and financial studies necessary for project planning and approval.

GREECE: Existing EIA arrangements include provision for an elementary type of environmental study in a number of regulations relating to urban development, forest protection, mining and quarrying, and protection of the marine environment; and a more formal system (as specified by Presidential Decrees 791 and 1180 of 1981) for licensing new industrial enterprises, which requires the submission of an EIA. However, these fragmented provisions are incomplete and it is aimed to replace them by a formal EIA system as an integral part of a new

Table 1. *continued*

environmental legal framework initially proposed in June 1983 and currently at the consultation stage.

IRELAND: There has been an administrative system for the environmental evaluation of new industrial projects since 1970, and a formal planning requirement for EIA in prescribed circumstances, since 1976. The legislation governing physical planning is contained in the Local Government (Planning and Development) Acts 1963 to 1983 and in accompanying regulations. However, both public participation, the coverage of studies and the types of project subject to EIA would need to be broadened for the European directive on EIA to be implemented. Relatively few statutory EIAs have been completed to date.

ITALY: Although existing regulatory provisions contain the embryonic elements of an EIA system, they are fragmented and incomplete and fall short of the basic formal requirements of such a system. In January 1984 an environmental bill was submitted to the national Parliament to introduce formal EIA procedures but it has not yet been enacted. Regional measures to introduce some elements of an EIA system have been introduced, or are proposed, in Lombardy, Emilia Romagna and Umbria.

LUXEMBOURG: In the late 1970s EIA procedures were introduced in Luxembourg law, through the Law of 17 July 1978 relating to the protection of the natural environment and Law of 16 April 1979 relating to dangerous, dirty and noxious installations. In addition, elements of environmental impact assessment may be incorporated into land use planning through the requirement for municipalities to prepare amenity plans (1937 Law) and through requirements relating to the general management of land use (1974 Law). 5–20 projects are believed to be submitted to some form of environmental evaluation each year.

THE NETHERLANDS: At the present, some EIA elements are included in various permit and physical planning procedures. Proposals for the introduction of a formal EIA system are contained in the 'Governmental Standpoint on Environmental Impact Assessment' (1979) and, in 1985, an environmental impact assessment act was passed. It is envisaged that 10–15 EIAs per year will be undertaken during the initial period of its application. A number of EIS studies are already being prepared, within the framework of an interim policy, and an extensive programme of EIA-related research is being undertaken.

UNITED KINGDOM: Although it does not have a formal EIA system there is now fairly widespread acceptance of the concept of EIA and numerous non-mandatory EIA studies have been undertaken mainly within the framework of existing development control legislation. The formal European Directive procedures can be integrated into the planning system with relatively little difficulty (the relevant legislation being the Town and Country Planning Act 1971), though the scope of planning studies and formal public participation may need to be broadened.

environmental legislation was being undertaken. Finally, the Directive was approved in June 1985. There is, of course, nothing to prevent member states from instituting EIA systems far more comprehensive and rigorous than the residual provisions put forward by the Commission. Many countries are making preparations to do so, or have done so, already.

Environmental Impact Assessment in the European Member States

Prior to the introduction of Community-wide assessment legislation, there was considerable development in the provisions made or planned for EIA by the (then) ten member states of the EC. These reinforced or extended the more embryonic elements of EIA systems which already existed by the early 1970s. The situation in the mid-1980s is summarized in table 1 (Lee, Wood and Gazidellis, 1984; Lee and Wood, 1985).

EIA legislation exists, or is proposed, in seven member states. In Germany, a form of environmental impact assessment is operated within the framework of a cabinet resolution. In Denmark and the United Kingdom, EIA-type studies are being undertaken to an increasing degree, within the framework of existing land-use planning legislation.[4] Both the scope and the detailed arrangements for EIA differ between member states. However, many similarities exist in the basic features of these arrangements and these are identified in table 2 (Lee and Wood, 1985). These common elements, which arise from both mandatory and non-mandatory EIA systems, bear a very strong resemblance to the requirements of the published draft Directive, which in turn incorporated the most important elements of a comprehensive EIA system. While tables 1 and 2 do not reveal the extent to which the systems in the different member states match the comprehensive model in practice, they do indicate that useful bases for EIA systems already exist.

The Commission of the European Communities is presumably hoping that, once a rudimentary mandatory system is in place, its benefits will become apparent as they have in the United States (Council on Environmental Quality, 1982) and that it may then become possible to strengthen it. Until then, of course, both the harmonization and environmental policy objectives of EIA will be only partially fulfilled.

Attitudes to Environmental Impact Assessment in the United Kingdom

There has been considerable official interest in EIA in the United Kingdom for several years. Several reports have recommended the acceptance of an EIA system (for example, Catlow and Thirlwall, 1976; Clark et al., 1981) but governments of both parties have been very cautious in their attitude to it. The House of Lords Select Committee set up to examine the draft Directive came

Table 2. Basic features of an EIA system.

EIA is mainly limited to projects, drawn from both private and public sectors, likely to have significant environmental impacts, although the longer-term trend is to extend the same kind of approach to other types of actions (for example, plans and programmes).

The *environmental impact studies* (sometimes called environmental impact statements) which are prepared for individual projects will normally be expected to cover the following types of items:
- description of the main characteristics of the project;
- estimation of residues and wastes that it is likely to create;
- analysis of the aspects of the environment likely to be significantly affected by the project;
- description of the measures envisaged to reduce harmful effects (this may be extended to include a consideration of alternatives to the proposed project and the reasons why they were rejected);
- assessment of compatibility of project with environmental regulations and land-use plans;
- non-technical summary of the total assessment.

The main *procedural* elements of the EIA process will normally include the following:
- the developer (often with assistance from consultants, regulatory bodies and other organizations) prepares an environmental impact study which is submitted, along with his application for project authorization, to the competent authority;
- the study is published (possibly after checking its adequacy) and is used as a basis for consultation involving both statutory authorities possessing relevant environmental responsibilities, and the general public;
- the findings of the consultation process are presented to the competent authority;
- the assessment study and consultation findings accompany the proposed project through the remainder of the competent authority's authorization procedure.

In a number of cases, these basic features are further elaborated, for example, by making arrangements for the preliminary *screening* of projects, for *scoping* the coverage of studies, for *independent panels* to vet the studies made for major projects, and for *monitoring* the environmental consequences arising from the implementation of the project.

down firmly in its favour after hearing evidence from a wide variety of bodies, including the Royal Town Planning Institute, in 1981:

> The Committee believe that the undertaking of assessments along the lines of the draft directive for major projects could perform a valuable function in ensuring that planning authorities in all Member States take proper account of the implications for the environment of proposed projects. (House of Lords, 1981*a*, pp. xxv–xxvi)

The government resisted the Committee's recommendations, an unusual occurence:

> The Government do not believe that the present draft directive yet gets it right. As a first step in a new field we consider it to be over-ambitious and likely to fail in its intention. (House of Lords, 1981*b*)

It is difficult to avoid the impression that the government was only paying lip service to the principle of an acceptable directive on EIA. Its commitment to environmental assessment basically amounted to leaving those planning authorities or developers who wished to carry out an EIA to do so. Any recommendations for any formalization of the system, whether emanating from the European Commission or not, had consistently been opposed (Wood, 1982).

The British position was that the European requirements for environmental impact assessment might duplicate or complicate current planning procedures. It was to ensure minimum disruption to the town and country planning system that the government insisted on major concessions as the price of its acceptance of the draft Directive. It seems probable that the House of Lords Committee report, as well as the concessions obtained, facilitated the eventual change in the government's attitude to EIA. The antagonism of British industry to environmental impact assessment (together with the enthusiasm of environmental groups for it) appears largely to have evaporated.

One manifestation of the official change of attitude is to be found in the Department of the Environment's draft code of practice for the pre-inquiry stage of major inquiries in which an 'environmental assessment (i.e study of the environmental implications)' was referred to (Department of the Environment, 1984). Here it was clearly anticipated that, even in the absence of mandatory requirements, many developers would have prepared EIA studies.

A further manifestation was the setting up, by the Department of Environment, of a Working Party, consisting of various departments and the professional bodies to study the issues surrounding the implementation of the Directive before it was approved. As well as examining the repercussions on planning practice, the working party also investigated the ramifications of the Directive on procedures where new projects were not subject to planning control (e.g. new highways). The DoE's Working Party reported in draft form for consultation in April 1986 (Department of the Environment, 1986). The final version is now expected in 1988. A further consultation paper was issued in early 1988 (DoE, 1988) and the final regulations were scheduled for summer 1988.

Impact of Environmental Assessment on British Planning Practice

The European Communities Directive is likely to have significant implications for planning practice. First, the vast majority of its provisions will be implemented within the town and country planning system in the United Kingdom and the 'competent authority' will be the local planning authority. Second, depending on the approach adopted by the Department of the Environment, there could be a considerable expansion of the scope of land-use controls. Finally, the operation of the planning system will need modification.

The incorporation of EIA in the planning system is unlikely to prove controversial. It appears that amendment of the Town and Country Planning Act 1971 will not be necessary to implement the Directive in the United Kingdom, but that changes to statutory instruments (e.g. the Town and Country Planning General Development Order, 1977) will be required.

The published draft Directive could have extended the scope of planning control in three ways: by applying the provisions of EIA to agriculture and forestry, both of which are currently outside planning control; by including modifications to projects which would frequently not require planning permission under the present system; and by granting the competent authority the right to alter conditions imposed on the development. The approved version of the Directive is not likely to be so sweeping in its effects. Much will depend on the way in which the Department of the Environment chooses to implement its provisions. It could still affect agriculture and forestry and include modifications, which currently escape the planning system, if the government chose to allow it to do so. The possibility of altering the conditions attached to a permission if a development proves environmentally unacceptable appears to have disappeared (this is already provided for under the Town and Country Planning Act, 1971, on payment of compensation).

There has been, of course, considerable recent debate about whether agricultural practices such as ploughing moorland for arable use or draining wetlands should be subject to planning control. The question whether or not large-scale afforestation should require planning permission has been debated for years. Both actions fall within Annex 2 of the European Directive and could consequently be subject to EIA if the Department of the Environment so determined (perhaps subject to certain criteria or thresholds) (DoE, 1988).

The acceptance of the draft Directive could therefore have involved a fundamental and controversial widening of planning powers. Whether or not local planning authorities would have been equipped to deal with such an increase in their responsibilities is debatable.

Intensification of use and changes of use within particular classes of the Town and Country Planning (Use Classes) Order (1984) have been identified as two of the most serious sources of pollution problems and of potential hazards (Miller and Wood, 1983). Such 'development' is normally permitted by the General Development Order and consequently falls outside the control of the local planning authority. The Directive, by including within Annex 2 'modifications to development projects included in Annex 1' and by naming specific projects, could substantially expand the scope of planning control. Many planning authorities would see this as a welcome means of assessing and preventing environmental problems that previously could only be alleviated by the payment of compensation under the planning acts, when action under pollution control legislation had proved ineffective. Once again, much will depend on the way the Department of the Environment chooses to implement this Annex.

The modification of the operation of the planning system to accommodate EIA involves several elements. It is useful to pick out a number of points relating to public sector developments: the provision of information by the developer; grounds for granting of planning permission; alternatives; public participation; and the grant of outline planning permission. The Directive seeks to apply the same provisions to public as to private developments. While significant procedural advances have been made in recent years (for example, the Leitch (1978) proposals on the assessment of trunk roads are now being implemented), there is still widespread concern about the way in which major public sector projects are put forward. The government is involved as both applicant and decision-maker. The various provisions of the Directive strengthen the procedural requirements (as opposed to conventions) involved and could lead to more careful scrutiny of the environmental impacts of public sector projects.

The information provided by the developer will form the basis of an assessment by the local authority. While there is nothing to prevent an authority from requesting (or requiring) that such information be produced at present, the requirement of the Directive ensures that this will be forthcoming. There is a need, however, to confine discussion to the key environmental issues and agreement between the developer and the local planning authority as to the nature of these might be necessary.

At present, except where the Secretary of State, or an inspector, makes a planning decision, the reasons for granting a planning permission are not stated. The Directive requires that reasons should be furnished in all cases where permission is granted following an EIA ('where the Member States' legislation so provides'). This runs counter to the fundamental assumption in the United Kingdom system that permission will be granted unless there are good reasons for its refusal. However, since EIA will only be applied to the more important developments, such reasons should not be difficult to express.

Annex 3 of the approved Directive mentions alternatives. The question of information about alternatives is a vexed one. The number of alternatives still feasible by the time the planning application is submitted may be very limited but a requirement to produce detailed data for each could result in increased costs, delays and voluminous documentation. This type of provision, however, has not proved overly onerous in the United States (Council on Environmental Quality, 1982). It would appear, therefore, that in practice it should not prove too difficult to strike a balance between the need for a reasoned justification and the costs of providing detailed information about alternatives.

The provisions for public participation in the Directive, while more demanding than current statutory provisions for public participation, fall short of those normally exercised at public inquiries in the United Kingdom. While unnecessary delays should obviously be avoided, it is only equitable for the public to see the competent authority's comments or assessment, together with the comments made by the administrative authorities and bodies responsible for environmental matters, before the final decision is reached. This procedure, which is not specified in the Directive, would be analogous to the plan-making process in Britain. It is, of course, open to the member states to decide that the participation requirements should exceed the Directive's requirements.

Once outline planning permission has been granted in the United Kingdom it cannot be revoked without payment of compensation. Any EIA for a major development must therefore be carried out before outline permission is granted. However, the information on which to base this is frequently inadequate and sufficient data often only become available at the more detailed stage of design, when the project may change significantly. There are two possibilities for improving the situation. The first is a requirement for a planning permission to be worded to take account of the fact that an EIA had to be completed (the EIA might be staged in such cases). The second would be to abolish outline planning applications where an EIA was required (as with change of use applications). In either case full and informal discussion between the local planning authority and the developer prior to submission of the application would be desirable.

It is not uncommon to employ EIA in current planning practice. By 1982 some 200 'environmental impact asssessments' (not necessarily meeting all the European Commission's criteria) had already been carried out in the United Kingdom (Petts and Hills, 1982). Most planning authorities undertaking these have been well pleased with the results and have certainly not experienced delays in determining applications. The costs (while difficult to determine) appear not to have been exorbitant, normally being less than 0.5 per cent of project costs (House of Lords, 1981a). Current practice suggests, therefore, that any mandatory system of EIA might be extended informally to unspecified projects by authorities determined to consider carefully the possible impacts. British expertise to carry out EIAs undoubtedly exists, providing numerous assessments are not required in the early years following the implementation of the Directive.

Conclusions

The detailed operation of the EIA provisions will depend upon the attitude and decisions of the Department of the Environment, since the Directive leaves a great deal to the discretion of the member states. It is probable that the United Kingdom central government will choose minimal compliance with the approved Directive; that is, very little change from the present system. Whilst such a change would be very significant in terms of prior consideration of environmental impacts and the release of information and consultation, it would fall short of environmentalist aspirations.

However, it seems likely that even minimal compliance with the provisions of the approved Directive would substantially boost the existing trend for local planning authorities to seek voluntary EIAs for environmentally significant developments. Wider use of existing best practice must lead to better overall consideration of environmental impacts, and to strenuous attempts to mitigate them. In terms of revenue and employment the temptation of any local planning authority is to accept new development. Consequently, any system which shows when a development would be environmentally unacceptable, and conspicuously reduces its adverse environmental effects, is obviously desirable.

The probability, therefore, is that local planning authorities would seek to extend EIA to developments not prescribed by central government. They would, in practice, seek to introduce a fairly extensive non-mandatory EIA system in addition to a limited mandatory one.

All in all, although the provisions of the Directive may fall short of the original hopes of environmentalists, they do represent real environmental gains. Yet the basic problems of planning control over environmental impacts remain. The use of EIA cannot eliminate the necessity of having to take a decision at one moment in time which, however carefully considered, will affect the environment for decades to come. Use of EIA can, however, ensure that the decision really is carefully considered and, most importantly, it is seen to be carefully considered.

There remain several unsatisfactory elements in the Directive, apart from the weakness of some of its provisions. In particular, the difference between the requirements specified in the main text and those suggested in Annex 3 is substantial. It is quite likely, of course, that there will be a tendency for local planning authorities to specify their requirements using Annex 3. This would be a further non-mandatory extension to the EIA system. There will be a real need for the publication of guidelines on the American model, by both the Commission and the Department of the Environment to ensure consistent implementation. Information exchange and training will also be required.[5]

The provision of more information, the necessity for increased consultation, and the compulsion actually to examine environmental impacts in detail seem certain to improve the environmental quality of decisions currently taken within the planning system. The scope of planning control might be significantly broadened, not least to major agricultural and forestry projects, again giving greater weight to environmental considerations.

If prevention really is better than cure, then better anticipation of environmental impacts can only be welcomed. Decisions may not be reversed, economic factors may still outweigh environmental factors, but at least the ramifications of decisions will be clearer. If United Kingdom experience at public inquiries is a guide (as it should be), some proposals are likely to be aborted when their impacts become evident. Many others will also be extensively modified, rendering them environmentally much more acceptable. This has certainly been the experience in the United States, where opinion about the benefits of EIA is now almost unanimously favourable.

NOTES

1. There are numerous books dealing with US procedures, with the various so-called comprehensive methodologies and with the assessment of particular impacts (see, for instance, Canter, 1977; Rau and Wooten, 1980). An excellent general introduction to EIA is provided by Clark, Bisset and Wathern (1980).

2. A few very minor modifications were made to the draft directive and published (CEC, 1982).

3. See, for example, Lee and Wood (1978). The contents specified in the draft Directive bore a strong relationship to the American and Canadian systems, for example.

4. For a comprehensive account of the provisions made for EIA in the ten member states and in various other countries see Lee, Wood and Gazidellis (1985).

5. For a fuller discussion about the need for information exchange and training in EIA see Lee and Wood (1985). Wood and Gazidellis (1985) provide a great deal of information relevant to EIA training and work experience, and Wood (1985) discusses the EIA training situation in the United States.

REFERENCES

Canter, L. (1977) *Environmental Impact Assessment*. London: McGraw-Hill.

Catlow, J. and Thirlwall, C.G. (1976) *Environmental Impact Analysis*. Research Report 11. London: Department of Environment.

Clark, B.D., Bisset, R. and Wathern, P. (1980) *Environmental Impact Assessment: a Bibliography with Abstracts*. London: Mansell.

Clark, B.D., Chapman, K., Bisset, R., Wathern, P. and Barrett, M. (1981) *A Manual for the Assessment of Major Development Projects*. Department of Environment Research Report 13. London: HMSO.

Commission of the European Communities (1977) European Community Policy and Action Programme on the Environment for 1977–1981. *Official Journal*, C139, 13 June 1977.

Commission of the European Communities (1979) *State of the Environment: Second Report*. Brussels: CEC.

Commission of the European Communities (1980) Proposal for a Council Directive concerning the assessment of the environmental effects of certain public and private projects. *Official Journal*, C169, 9 July 1980.

Commission of the European Communities (1982) Proposal to amend the proposal for a Council Directive concerning the assessment of the environmental effects of certain public and private projects. *Official Journal*, C110, 1 May 1982.

Commission of the European Communities (1985) Council Directive of 27 June 1985 on the assessment of the effects of certain public and private projects on the environment. *Official Journal*, L175, 5 July 1982.

Council on Environmental Quality (1982) Implementation of CEQ regulations on NEPA. Document dated 12 July 1982. Washington, DC: CEQ.

Department of the Environment (1984) Code of Practice for the Pre-inquiry Stages of Major Inquiries. PLUP1/48/16. London: DOE.

Department of the Environment (1986) Implementation of the European Directive on Environmental Assessment. Consultation Paper. Department of Environment. London. Mimeo.

Department of the Environment (1988) Environmental Assessment. Implementation of EC Directive. Consultation Paper. Department of Environment. London. Mimeo.

House of Lords (1981a) *Environmental Assessment of Projects*. Select Committee on the European Communities, 11th Report, Session 1980–81. London: HMSO.

House of Lords (1981b) EEC 11th Report: Environment. *Parliamentary Debates Official Report, Session 1980–81*, 30 April 1981, pp. 1311–47.

Lee, N. and Wood, C. (1978) EIA – a European perspective. *Built Environment*, 4(2), pp. 101–10.

Lee, N. and Wood, C. (1985) Training for environmental impact assessment within the European Economic Community. *Journal of Environmental Management*, **21**, pp. 271–86.

Lee, N., Wood, C.M. and Gazidellis, V. (1984) *Training for Environmental Impact Assessment*. Directorate General for the Environment Consumer Protection and Nuclear Safety, Commission of the European Communities, Brussels.

Lee, N., Wood, C. and Gazidellis, V. (1985) *Arrangements for Environmental Impact Assessment and their Training Implications in the European Communities and North America: Country Studies*. Occasional Paper 13, Department of Town and Country Planning. Manchester: University of Manchester.

Leitch, G. (1978) *Report of the Advisory Committee on Trunk Road Assessment*. Department of Transport. London: HMSO.

Miller, C. and Wood, C. (1983) *Planning and Pollution*. Oxford: Oxford University Press.

Petts, J. and Hills, P. (1982) *Environmental Assessment in the United Kingdom*. Institute of Planning Studies. Nottingham: University of Nottingham.

Rau, J.G. and Wooten, D.C. (eds.) (1980) *Environmental Impact Analysis Handbook*, London: McGraw-Hill.

Town and Country Planning General Development Order 1977. Statutory Instrument 1977, No. 289, as amended, London: HMSO.

Town and Country Planning (Use Classes) Order 1987. Statutory Instrument 1987, No. 764. London: HMSO.

Wandesford-Smith, G. (1979) Environmental impact assessment in the European Community. *Zeitschrift fur Umwelt Politik*, 1, pp. 35–76.

Wood, C. (1982) The impact of the European Commission's directive on environmental planning in the United Kingdom. *Planning Outlook*, **24**, pp. 92–98.

Wood, C. (1985) The adequacy of training for EIA in the United States of America. *Environmental Impact Assessment Review*, **5**, pp. 321–37.

Wood, C. and Gazidellis, V. (1985) *A Guide to Training Materials for Environmental Impact Assessment*. Occasional Paper 14, Department of Town and Country Planning. Manchester: University of Manchester.

Chapter Six

Assessing the Environmental Impacts of Policy

PETER WATHERN, IAN BROWN,
DAWN ROBERTS, and STEVEN YOUNG

Recent major planning inquiries in the United Kingdom such as the Drumbuie platform fabrication yard and the Windscale reprocessing plant proposals (Clark, Bisset and Wathern, 1981) as well as the recent Sizewell pressurized water nuclear reactor project, have revealed the links between strategic development proposals and the national policies which underlie them. In particular, the Windscale and Sizewell inquiries have revealed the problems of attempting to accommodate consideration of both future national energy policy and detailed matters of land-use planning.

Paradoxically many critical issues relating to major development proposals may be pre-empted during the formulation of a policy. Similarly, a series of individual development projects may dictate the nature of a national policy by forcing it inexorably in a particular direction. A rational planning system should ensure that the environmental, social and economic consequences of a policy are considered during its formulation, whilst the implications of individual proposals are assessed adequately before any decision to proceed.

This chapter indicates how the implications of a general policy might be assessed. It describes research into methodologies for assessing the environmental impact of policies undertaken for the Commission of the European Communities. In order to develop an approach of general application, research has focused on non-environmental as well as environmental EEC policies and even United Kingdom national policy has been considered. Throughout the discussion environmental impact assessment (EIA) is used in its widely accepted planning context, namely the assessment of individual development projects. The term 'policy appraisal' is used in connection with the assessment of policies. O'Riordan and Sewell (1981) have used the alternative phrase 'policy review' to describe this process.

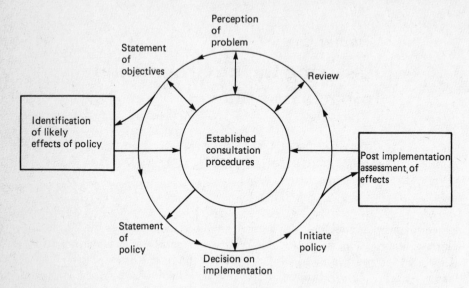

Figure 1. Technical inputs to policy formulation and review.

Approaches to Policy Appraisal

EIA with respect to development projects has been accepted throughout the world not only because of its obvious immediate benefits, but also because of the ease with which it can be implemented within existing procedures and practice. Lee (1983) has indicated that the extension of appraisal is a key priority for future environmental research 'to improve understanding of specific areas of policy . . . assessment'. An important outcome would be:

> to raise the general level of confidence in its practical usefulness and make it easier to extend its application into the earlier stages of policy . . . formulation where its hold is relatively weak at present'.

The inability to isolate the causes of environmental change and to ascribe them to particular policies has been the principal factor which has precluded the development of a policy appraisal technique in the past. Unless the effects of a policy on the environment can be identified and isolated from other causes of change, it is impossible to assess either its impacts or its effectiveness in protecting environmental quality. To be an effective tool in environmental management, however, policy appraisal should not be merely a retrospective review of the impacts of policies after implementation. Of greater significance is the need to assess the possible environmental implications of a policy during its formulation.

The need for a systematic procedure is greatest in the case of non-environmental policies. Environmental policies would not normally be expected to cause environmental degradation, and policy appraisal is likely to focus on ascertaining whether they would achieve their objectives. Non-environmental policies, however, are more likely to result in unexpected environmental damage. It is only through the early identification of such possible conflicts that environmental management can become fully comprehensive in its scope.

Although there are similarities between project-based EIA and policy appraisal, the geographic area affected by a policy is greater, and its impacts more diffuse than with a specific development project. There is an abundance of literature on project-based EIA but there are few descriptions of the objectives and information requirements of policy appraisal (Clark, Bisset and Wathern, 1980). The appraisal of plans, however, is closely allied to policy appraisal but the literature on this subject is also quite limited, and focuses largely on the difficulties of its integration into existing administrative procedures (see, for example, US Department of Housing and Urban Development, 1977; In't Anker and Burggraaff, 1979; Environmental Resources Ltd, 1981; De Jongh, 1982).

The application of EIA in plan-making and the method relevant at various stages in the planning process have been identified (Wood, 1982; Lee and Wood, 1980). An inventory approach to plan assessment may be appropriate (Hall, 1977). Methodological development in plan assessment is generally poor and existing EIA methods remain largely untested in this context (Wood, 1982); thus little experience of project-based EIA can be transferred to the consideration of policies.

Before an appropriate method can be developed it is important to identify the criteria required for policy appraisal. First, it is necessary to identify the effects of a policy in advance of its implementation. Second, once a policy has been implemented it must be possible to isolate environmental changes attributable solely to its influence. Finally, the perceived changes in environmental quality must be related to the predicted effects in order to assess the effectiveness of policy.

It is important to reduce both the information requirements and the complexity of appraisal to manageable levels. There would be considerable methodological difficulties in trying to isolate every environmental change. Such an analysis would require extensive environmental quality monitoring over a wide geographical area. Data requirements would be reduced and the analysis less complex if appraisal were confined to a small number of pre-selected areas believed to be 'sensitive' to the proposed policy changes.

The methods currently used in project-based EIA might be more generally applicable in policy appraisal: for instance, predicting the impacts of policy and isolating its effects after implementation become analogous to a project EIA and its post-development audit although the structure of the appraisal method itself is, of necessity, more complex.

A statement of the objectives which a policy should satisfy is required at an early stage in order to identify alternative measures which could achieve similar objectives. Systematic consideration of the possible environmental impacts of each alternative should ensure that proposals likely to cause unnecessary environmental degradation are rejected at the outset. Subsequent appraisal activities are associated with the selection of suitable monitoring sites and the development of programmes to generate appropriate data for isolating the effects of the policy after implementation.

This chapter contains only a summary of the policy appraisal process (for a more detailed account see Wathern et al., 1983a). The stages involved in the development of a policy and the points of contact between existing administrative procedures and policy appraisal are shown in figure 1. The figure indicates that two important technical aspects of policy appraisal are the assessment of the potential impacts of a proposed policy and the post-implementation review of its effects. The following discussion is therefore concerned with:

(a) assessment of the potential impacts of a proposed policy;
(b) implementation of policy;
(c) post implementation assessment of policy effects.

Assessment of the Potential Impacts of a Proposed Policy

The technical procedures involved in policy formulation are illustrated in figure 2. These should be regarded as components of an iterative decision process, because each is likely to be involved repeatedly during the formulation of a particular policy. Information about the environmental aspects of a policy is generated at several points in the decision process. This information can be supplied via the established administrative and consultation procedures. There need be no formalized end point to these steps although some form of report is likely to be produced. Four activities are involved in assessing the implication of any given policy:

(a) establish baseline situation;
(b) identify likely impacts;
(c) identify appropriate monitoring scheme;
(d) integrate with other policies.

(a) Establish Baseline Situation

There are three types of EC policy which have significant environmental implications. First, 'reductionist policies' are formulated to reduce either critical levels of pollutants or rates of change in the environment, for example the

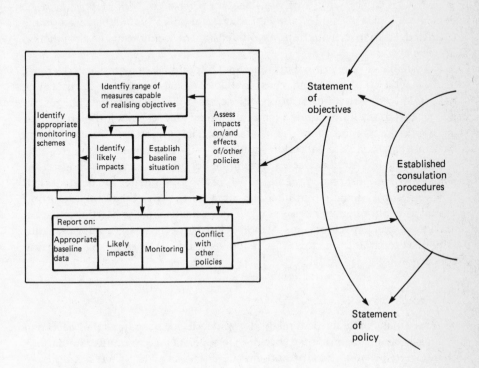

Figure 2. Procedures for determining likely impacts of a policy.

Directive on smoke and SO_2 levels (Council of the European Communities, 1980). Second, 'protectionist policies' are formulated to protect non-stressed sectors of the environment which comply with prescribed standards, and to improve those sectors which should comply with the standards, for example the EC freshwater fish Directive (Council of the European Communities, 1978). Finally, 'non-environmental policies' may cause change in environmental quality as an indirect consequence of their implementation, such as the Less Favoured Areas (LFA) Directive (Council of the European Communities, 1975).

This division describes most types of policy and suggests what criteria should be used for assessing their impact. Thus, the effectiveness of protectionist policies can be tested in the maintenance of high quality resources, or the effectiveness of reductionist policies can be tested by the extent to which grossly degraded resources have been improved.

Some policies, such as the Dangerous Substances Directive (Council of the European Communities, 1976) contain more than one objective and are hybrid in nature. Clearly with such policies, the impact of all facets should be evaluated. Site selection for non-environmental policy should be based on the predicted spatial distribution and intensity of significant environmental impacts.

Knowledge of the pre-policy (baseline) situation is a necessary reference against which to assess changes in environmental quality. A thorough description of the baseline situation will increase the likelihood that changes can be ascribed to the influence of a particular policy. It is essential, therefore, that an appraisal method should include adequate provisions for assembling baseline data, and data collection should be initiated at an early stage.

The sites should not be selected at random, but positioned where the effects of a policy will be most readily manifest. Before appropriate sites can be identified, however, some prediction of likely impacts, particularly their magnitude and spatial distribution, may be necessary. Thus, establishing the baseline situation and identifying impacts are closely inter-related aspects of policy appraisal and, although separated in this discussion, should be viewed as occurring concurrently.

(b) Identify Likely Impacts

There are many available EIA methods for identifying the potential impacts of a proposal, ranging from simple checklists to sophisticated computer simulations of anticipated change. EIA methods are not discussed in detail here as they have been reviewed adequately in many different publications, for example Clark, Bisset and Wathern (1980), Bisset (1984) and Wathern (1984a,b). There are, however, a few methods which appear to be particularly appropriate for policy appraisal.

Matrices and networks have gained widespread acceptance in the assessment of development projects mainly because of their relative simplicity. Bisset (1984) considers that the Leopold matrix (Leopold et al., 1971) has been most widely adopted. The main criticism of the approach, however, is that it fails to take account of indirect, secondary and higher order effects, a feature which may limit its utility in policy appraisal.

Environment Canada (1974) has adopted the component interaction matrix to resolve this problem. A mathematical technique, matrix algebra, is used to detect indirect impacts, based only upon consideration of the direct effects. Wathern (1984a,b,) reports the incorporation of quantitative data into the method. Whilst greatly extending its power as an analytical tool, this has greatly increased its information requirements. The method is not yet widely known, even in its qualitative form.

The second type of method which accommodates indirect impacts is the network approach such as Sorensen (1971). These methods use an analysis of known networks to explore the ramifications of particular activities upon a system. There are two serious practical difficulties in using networks. They are cumbersome and, as the analysis is hierarchical, are ill-equipped to accommodate feedbacks and cycles within a system.

An interactive computer-based network approach has been developed for the US Forest Service (Thor *et al.*, 1978). A computer-based expert system would be particularly useful when there are constraints on either the time available for consultations, or appropriate expertise. Both factors are likely to be significant constraints during policy formulation. The advantage of a computer-based approach is that the components of the environmental system, with all its inherent complexities, are subsumed within the computer software and there is no need for the operator to understand the interactions involved. It seems that a method for identifying the impacts of a particular policy could be constructed readily from these three components, namely simple matrices, component interaction matrices and networks.

(c) Identify Appropriate Monitoring Schemes

Policy appraisal should include the means for assessing the effects of a policy, particularly the necessary monitoring programme. Hitherto, monitoring has either been ignored or treated as an *ad hoc* postscript to policy assessment. The inadequacy of present procedures is demonstrated, for example, by the EEC bathing beach Directive (Council of the European Communities, 1976) which indicates that fortnightly sampling throughout the bathing season is acceptable for the purposes of estimating the 95 percentile *E. coli* concentration in bathing water. However, the degree of uncertainty in estimating such a parameter from so few samples renders the exercise '. . . virtually useless . . .' (Ellis and Lacey, 1980).

One dilemma which has faced environmental managers in the past is to determine at what stage in assessment monitoring schemes should be established. As the function of monitoring is to provide data on the situation after implementation of a policy, for comparison with the baseline situation, it should be established at an early stage. If data collection is delayed, truly baseline data will not be available, and no amount of trend data can remedy this deficiency. The major disadvantage of early implementation of data collection schemes is the risk that a policy will not be initiated, resulting in the data becoming redundant. Although this risk is real, it is perhaps overstated, because the data may have already served an important function by influencing the decision not to proceed with a policy.

The close link between baseline studies and monitoring schemes has been stressed in the literature. For example, the US Environmental Protection Agency (1976) in its Ocean Dumping Regulations describes baselines as 'trend assessment survey'. Thus, baseline studies are concerned with establishing the dynamics of change within a system prior to implementation, whilst monitoring is concerned with the dynamics after implementation.

In the static system, the effectiveness and impacts of a policy are assessed by comparing the baseline state with that after implementation, whilst in dynamic systems rates of change should be compared. The intensity of monitoring required for a particular policy, however, is related to the level of uncertainty that exists within the system under consideration. Three factors influence the level of monitoring that must be established: the nature of the policy under review; the stability of the environmental system being considered; and the inherent variability in the measured parameter.

For example, the effect of policy on the survival of certain adult raptors, would be easy to demonstrate. Peregrine falcons (*Falco peregrinus*) nest at traditional sites and the total carrying capacity of a particular region can be determined. The number is stable. The effect on overall population levels could be assessed from limited observations on the number of pairs attempting to breed at a small subset of the total nest sites. Other parameters of the species, such as successful production of young from nests, or the rate of egg production per nest, are less stable. They may be greatly affected by natural factors in the environment, and by density dependent factors within the population. Thus, in the case of peregrines, percentage egg production has ranged from 4.5 to 35 per cent over the period 1961–79, while territory occupation has varied from 18 to 58 per cent, not always in concert. Red kite (*Milvus milvus*) populations have been rising from 1946 to the present. Although the percentage of territories successfully occupied has remained unchanged, the percentage rate of young production has declined slightly from 34 to 27 per cent. Thus, policies affecting naturally unstable aspects of their breeding biology would be detected only with a more intensive programme of monitoring over a longer period than a policy causing the demise of adult birds.

There may be special problems associated with those instances where it is impossible to collect sufficient data on the baseline situation. Occasionally, this can be accommodated by selecting a comparable area which is likely to remain unaffected by the proposal. This may be considered as a reference against which to assess the trends occurring at the monitoring site. This is not a strict scientific 'control', but it can be used to indicate natural change. For many natural changes with long periodicity, such sites may represent the only means of detecting non-seasonal cyclic change, and the role of such reference sites needs to be evaluated further. Their use may be limited to certain types of policy such as those controlling discharges from specified industries. The concept would not be applicable for policies with Community-wide or even inter-regional impact, because of the dramatic regional differences in the response of some systems to change. It is essential, therefore, that the reference site should lie within a region showing similar responses to the monitoring site. The ideal situation would be one in which the reference site could be built into the system under consideration, for example, a river having a single pollution source, or a sequence of sources which could be controlled by the policy.

The timing of an impact may have implications for monitoring, because some impacts are likely to be manifest shortly after implementation; others are more medium term, while some are likely to be experienced only in the long-term. The time-scale of the anticipated impact should determine when data from a specific site cease to be considered as representative of the baseline situation. Data collected immediately after implementation of a policy may be more indicative of baseline processes and rates of change than of the effects of the policy, particularly with respect to long-term impacts. These data may be used legitimately to improve the understanding of the baseline situation. The judicious use of resources is fostered by a careful analysis of the time-scale of change. There is little value in attempting to measure an impact before it is likely to be manifest. Many of the more subtle impacts of policy are likely to be manifest over a long period of time.

(d) Integration with Other Policies

The broad objectives of EC environmental policy are defined in a series of Action Programmes. In the most recent (the third), the need for the integration of all facets of Community policy is stressed (Council of the European Communities, 1983). This should ensure that policies emanating from different sectors of the Community are not in conflict. Where apparent conflict exists this should be the result of a conscious decision that, in one sphere of activity, the demands of a particular policy will take precedence over another, rather than a failure to detect conflict or inconsistency. The process of integration should also involve a consideration of the implications for, rather than merely the conflicts with, existing and proposed policies.

The importance of this facet of policy appraisal cannot be overstated. If a non-environmental policy, for example, an agricultural or regional policy, is being formulated the process of integration with other policies may represent the only opportunity for considering environmental aspects. This has certain implications for appraisal. There are likely to be severe time constraints on consultations concerning possible environmental effects, because this will be but a small component of the overall process. Thus, a policy appraisal system must be efficient and initiate early consultations.

It may be necessary to reconsider a policy in the light of its implications for other policies. An alternative may be equally effective while having reduced environmental impacts. Thus, for example, a policy for sustaining population levels and raising living standards in sparsely populated areas by grant aid for increased agricultural production could have an extensive impact on wildlife. This might be replaced by a policy of small-scale industrial or recreational development with intensive but localized effects. If the outcome of these deliberations is a decision to implement a policy, consideration should be given to the implementation procedures which should be adopted.

Implementation of Policy

The processes involved in implementation can be described most readily by consideration of EC policy, given the unequivocal statements of intent which exist and the clearly defined instruments which are available for its adoption. United Kingdom national policies are rarely stated so explicitly. The analysis, however, is confounded by the, often antagonistic, influences of individual Member States.

EEC environmental policy is stated in a series of Action Programmes which have been adopted at intervals since 1973. Various legislative means for implementing this policy are available to the Community, namely, 'Regulations', 'Directives', 'Decisions', 'Recommendations' and 'Opinions', although the latter two are not binding upon Member States. Directives have become the major vehicle for the introduction of environmental policy, although much agricultural policy is implemented through regulations.

The most important EC environmental Directives relate to water quality, waste disposal, noise, air quality and wildlife. In addition some agricultural Directives such as the Less Favoured Areas Directive also affect the environment. Member States are required to introduce national legislation to accommodate such Community provisions where these are not already covered by existing statutes, and must formulate administrative procedures for implementation. Allowing Member States to formulate procedures to comply with a Directive has important implications in determining the effectiveness of a policy. These implications are discussed in detail in the next section.

Although individual Directives differ substantially, they have two main types of structure. Some Directives are very explicit, with detailed objectives and compliance dates specified within them. The smoke and SO_2 Directive and many of the water quality Directives have this structure. The remainder lack this specificity. Although their objectives are defined in general terms, establishment of the timetables for reduction of specific environmental pollutants and specific target values within these timetables is the responsibility of individual Member States. The Directive concerned with waste from the titanium dioxide industry (Council of the European Communities, 1978) is of this type.

Directives are formulated over a long period, involving discussions with all Member States. These discussions provide an opportunity for proposed objectives and criteria to be closely scrutinized and modified as appropriate. The adopted Directive is an agreed text acceptable to each Member State. Even when objectives are clearly defined, however, Member States may be given considerable freedom in the interpretation of crucial aspects of a Directive. Seemingly unassailable provisions can be reduced, subsequently, to mere administration burdens by Member States unsympathetic to the objectives of a particular piece of legislation. Thus, individual Member States frequently are provided with two opportunities to influence the potential effectiveness of a policy. The approach of an unsympathetic Member State can involve both pre-emasculation during the period of policy formulation and post-adoption deflection of intent.

Rarely do EC Directives effect instantaneous change as usually some future compliance date is set. The smoke and SO_2 Directive, for example, has two 'target' dates and these are separated by a decade. The effects of this policy, therefore, are likely to be phased over a prolonged period. In addition, action may be initiated in anticipation of a policy being implemented. Many United Kingdom Water Authorities have responded to EC water quality Directives in this way.

Policy appraisal, therefore, must accommodate the various procedures that may be adopted to implement a particular policy. When policies contain specific objectives and compliance dates, appraisal should be accomplished most readily. Policy which is loosely formulated, however, may be difficult to appraise. When a particular Member State has sought to deflect the intent of a policy, however, a lack of effectiveness may reflect either the characteristics of the policy or its mode of implementation.

Post-Implementation Assessment of Policy Effects

Wathern et al. (1983b, 1985a) have argued that there are three discrete phases involved in the assessment of the impacts of a policy after it has been implemented. These have been termed 'legislative', 'procedural' and 'substantive' review.

At an early stage in an appraisal it is important to establish that the appropriate statutes to implement a policy have been enacted. This is particularly important with international regulations such as EC Directives, as national governments must enact appropriate enabling legislation. The Commission currently undertakes such a legislative review with respect to each Member State to ensure that there is formal compliance with EEC provisions. Legislative review is not considered further in this discussion, as an admirable review of the relationship between EC environmental policy and United Kingdom domestic legislation is given by Haigh (1984).

The second phase of appraisal, procedural review, should seek to establish the mechanism by which a particular policy has been implemented. The main objective should be to establish the extent to which detailed implementation procedures facilitate or impede realization of policy objectives. For example, Wathern et al. (1983b) have shown how the administrative procedures for implementing the shellfish Directive in Wales have thwarted realization of its objectives.

In Wales, six commercially important shellfisheries were considered for designation, but only one in the Menai Straits has been designated. It alone complies with the water quality standards defined under the Directive. Other commercially more important, but environmentally more contaminated, fisheries are undesignated. Considerable capital expenditure would be required for sewage treatment to raise water quality in these fisheries if they were to be designated. This has effectively been ruled out by central government advice on capital expenditure. Thus the procedures have ensured that EEC regulations will not be allowed to intrude upon United Kingdom environmental policy (Young and Wathern, 1984).

A similar situation is revealed by a procedural review of the implementation of the Less Favoured Area (LFA) Directive in Wales (Wathern *et al.*, 1985*b*). The objectives of the Directives are to maintain a minimum population level in sparsely populated areas and to conserve the countryside in these regions. Implementation of the Directive, however, is subject to national interpretation. United Kingdom central government has neglected important objectives of the Directive in opting to use it to increase agricultural production in line with the long-standing policy towards the uplands embodied in the Hill Farming Act of 1946 and subsequent legislation.

Although Haigh (1984) argues that the Directive is essentially an agricultural policy, the narrow interpretation adopted within the United Kingdom has been criticized as being too narrow and restrictive (MacEwan and Sinclair, 1983; Countryside Commission, 1984). Indeed, Wathern *et al.* (1985*b*) have further argued that the provisions which have been adopted have probably made a significant contribution to the degradation of the environment in upland areas. This procedural review of the Directive, therefore, indicates that implementation within the United Kingdom embodies the potential for considerable adverse environmental impact.

Substantive review is dependent upon the ability to identify points in time when changes in environmental quality occur, and subsequently to identify actions responsible for these changes. The effects of a policy can be separated from the effects of other factors if the actions causing environmental change can be shown to have been taken as a consequence of the policy under review. Unless a policy is formulated in such a way that it will initiate, or actively prevent, certain activities, it will be impossible to isolate any resultant change. This consideration effectively limits the policies which can be appraised.

The activities involved in the assessment of a policy after it has been implemented are illustrated in figure 3. The objectives of such a review are to determine the quality of the environment before and after implementation, to isolate the effects of the policy, and to report the conclusions to the established administrative and consultative procedures. If information on the effects and effectiveness of a policy is made available when important adverse environmental changes are first detected, re-appraisal should be possible before irreparable damage is caused. In this way, policy appraisal should become an effective and sensitive tool in environmental management.

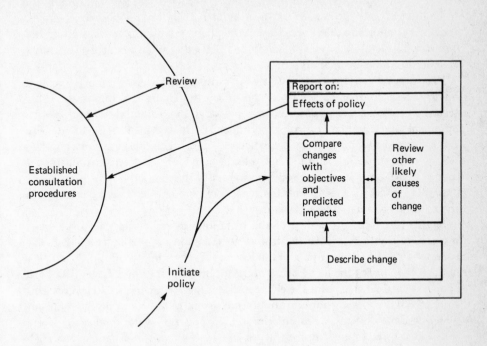

Figure 3. Procedures for reviewing effects of a policy.

There are three stages involved in isolating the effects of a particular policy on the environment. First, changes in environmental quality must be identified. Secondly, these changes must be linked to their causal factors. Finally, the causal factors, in turn, must be linked to the influence of a policy. By far the most subjective activity in policy appraisal is to determine whether any particular action causing environmental change was initiated simply because of the existence of a policy. The degree of subjectivity involved inevitably influences the decision as to whether attempts should be made to quantify the impact. Only when the investigator is sure that individual changes would not have been implemented without the existence of a particular policy, can effects be reasonably quantified. Alternatively, there will be instances when monitoring programmes can be designed and monitoring sites selected in such a way that the recorded changes are unquestionably a direct consequence of the policy under review. Unless either of these criteria can be satisfied, attempts to use a quantitative technique would be both wasteful and inaccurate.

The major difficulty in policy appraisal, dissociating real from random change, is a problem common to all aspects of environmental management. Although various statistical techniques have been used to appraise the impact of individual policies, no consensus on a standard analytical method has emerged. Indeed it is desirable that the nature of the problem under investigation should continue to dictate the type of technique used, at least until considerably more case studies have been analysed.

Previously, analytical methods used in industry for production management and process control have largely been ignored in policy appraisal. These techniques, however, appear to have considerable potential, as both policy appraisal and quality control in production management seek to identify significant shifts in a long sequence of data. Cumulative sum analysis (Woodward and Goldsmith, 1964) has been widely adopted in industry. It involves a sequential comparison of individual values with the overall mean of a data set in order to detect points in time when there are significant deviations in the parameter under consideration.

In this research, the utility of the technique in policy appraisal has been investigated with reference to air and water quality surveillance data. It has been used to assess the impact of the Rivers (Prevention of Pollution) Act 1961 upon the quality of the River Ebbw at the monitoring station at Cwm (Wathern et al., 1983a). At this point the effluent discharged from an industrial concern, situated 3 km above the monitoring station, is one of the major determinants of water quality. Policies for raising the standard of river water quality are likely to focus on effecting improvements in the effluents discharged by such industries. As the effluent has considerable reducing potential, changes in the 4-hour-permanganate values at Cwm since 1967 have been analysed in an attempt to detect the effect of the policy. Although techniques for assessing environmental quality before and after a particular action can indicate the significance of the changes, they cannot be used to identify the factors which may have caused them. Thus, the fall in average value of permanganate values at Cwm from 47.9 mg/l during 1967–70 to 6.2 mg/l during 1979–1982 is statistically significant (students' t-test). The intervening period, however, was characterized by widely fluctuating values, and the analysis gives no indication of the likely cause of improvement.

Using cumulative sum analysis, however, it has been possible to show that this improvement in river water quality at Cwm was a direct consequence of the installation of new treatment facilities at the works. The impetus for the new facility was the policy adopted by the Welsh Water Authority for improving the quality of Welsh rivers as required by the 1961 Act, even though the treatment plant was not commissioned until 16 years after implementation of the legislation.

The analysis performed on the River Ebbw using the cumulative sum technique can be invalidated on strictly statistical considerations, even though it gives interpretable results. Woodward and Goldsmith (1964) indicate that this technique requires samples to be taken at regular intervals. The River Ebbw data do not satisfy this criterion. For the purposes of the analysis, however, the data were treated as if collected at equal time intervals.

Trends in atmospheric quality, particularly smoke and SO_2 are amenable to analysis using the cumulative sum technique because data are collected on a regular basis. Monthly, seasonal and annual averages can be assessed either for individual sites or for aggregates of all sites within a particular category. Sites in the high-density housing category (Warren Spring, 1982) over the period 1962–82 were selected for detailed analysis. The major advantage of analysing trends in air quality using annual means is that seasonal variation is subsumed within the overall average and general trends can be detected more readily. Considerable detail, however, is sacrificed in the analysis. Paradoxically, splitting the data into winter and summer averages makes the analysis more difficult to interpret, mainly because of a drift in seasonality in successive years. The arbitrary division of the year into summer and winter months, April-September and October-March respectively, may not correspond to the real seasons in any one year.

The most satisfactory cumulative sum analysis was based upon SO_2 data from a single site, Newport. Initially, winter and summer means were analysed separately. An additional refinement was incorporated into the analysis to accommodate seasonality. Points of change in air quality were determined and the statistical significance of each change assessed. Summer and winter analyses were recombined into a single graphical presentation. Using this approach, changes in air quality could be related to such events as the international oil crisis, two miners' strikes and the onset of the economic recession. Clearly cumulative sum analysis has considerable potential in policy appraisal.

Occasionally, changes in practices affecting environmental quality may be linked with confidence to the requirements of a policy. For example, reductions in the sulphur content of heating oils in Montreal were unlikely to have occurred without the introduction of heating oil regulations. Similarly, petroleum companies were unlikely to have reduced the lead content of petrol in West Germany without the influence of Petrol Lead Laws. Roy and Pellerin (1982) and Jost and Sartorius (1979) respectively have attempted to quantify the environmental impacts of these policies. These attempts are clearly justified, as the criteria for quantification outlined above are satisfied.

Roy and Pellerin used intervention analysis to quantify the effects of heating oils regulations in Montreal on air quality. This policy established progressive reductions in the sulphur content of oils which could be sold for heating systems, and set three compliance dates in successive years. Intervention analysis requires the dates when environmental quality could be expected to change to be known. Roy and Pellerin used the three compliance dates to represent three separate 'interventions'. This decision, however, introduces crucial unrealistic assumptions. A degree of accuracy in the analysis was forfeit because the regulation sought to control the quality of heating oils at their supply point, not at the point of combustion. Almost certainly, oils of higher sulphur content were burned after each of the three compliance dates. Thus, the assumption that the policy would cause instantaneous change is not acceptable and the dates at which it became fully effective are probably not the stated compliance dates. The initial use of a technique, such as cumulative sum analysis, to identify the onset of changes in air quality would have improved the accuracy of this appraisal considerably.

Jost and Sartorius (1979), calculated average lead concentrations in urban streets before and after implementation of regulations reducing the permissible lead content of petrol to assess the impact of West Germany's Petrol Lead Laws. These averages were compared with similar data collected from largely traffic-free industrial zones. Reductions in ambient lead levels of around 60 per cent were recorded in streets with heavy traffic, whilst approximately 30 per cent reductions occurred in urban streets not exposed to heavy traffic. In industrialized areas, concentrations fell by approximately 20 per cent. Jost and Sartorius conclude that lead sources other than automobile exhausts can be neglected in heavy traffic streets, and that the observed reductions in these areas must be caused solely by the Petrol Lead Laws.

Frequently, specific chemical compounds are associated with particular polluting emissions. Analysis of changes in the concentrations of these specific compounds may indicate the influence of a particular policy. Newly discharged particulates from vehicle exhausts, for example, are composed primarily of lead:bromine (Pb:Br) compounds (Hopke et al., 1976). On ageing, bromine is lost from these compounds and combines with the ambient aerosol. Lead oxides (PbO_x), carbonates ($PbCO_3$, $(PbO)_2 PbCO_3$) and halides (PbBrC1) are left as particulates. If Jost and Sartorius had monitored the concentration of these compounds close to roadways rather than total lead, the analysis would have indicated the contribution of automobiles to urban lead levels and, consequently, a more accurate evaluation of the Petrol Lead Laws.

Should quantification of impacts be a significant objective of appraisal, selecting monitoring sites likely to be affected by the policy alone becomes of paramount importance. As more factors operate simultaneously in determining environmental quality at a particular site, separating and quantifying the influence of each becomes increasingly difficult. Using sites where the only major influence on environmental quality is the policy under review will result in a more accurate and straightforward analysis. The analysis of air quality at Newport, however, shows that stations with good quality environmental data bases frequently are complex, having been sited amidst many polluting influences.

In certain circumstances, multivariate statistical analysis could be used to accommodate such environmental complexity. Hopke et al. (1976), for example, used factor analysis and hierarchical aggregative cluster analysis to identify not only the most significant factors influencing Boston urban aerosol quality, but also the sites most susceptible to each of these factors. Such analyses could be used in site selection to identify those areas most sensitive to particular changes and least influenced by non-policy induced effects. This approach would have improved the analysis by Jost and Sartorius because only sites at which air quality was determined by vehicle emissions would have been selected, imparting greater cogency to the results.

Policies, however, may also affect the environment more qualitatively and the effectiveness of such policies may be difficult to assess. For example, in the post-war period there have been a number of local authority policies which have affected the sand dunes in North Clwyd. The impacts of these policies have been reviewed by Brown *et al.* (1985).

The value of this study lies in the ability not only to trace changes in the condition of the dunes over time, but also to relate these changes to an evolving policy for the area. The policy has changed from one favouring use of the dunes for recreation to a conservation policy. The change has been a phased one. Analysis of aerial photographs indicates clearly that the period up to the early 1970s was one of dune degredation. This was followed by a period up to 1974 when parts of the dunes improved whilst other areas continued to degenerate. Since that time there has been continued improvement, apart from one area close to the Point of Ayr.

These vegetational changes can be interpreted in terms of policy. The local authority in its Development Plan of 1954 determined in favour of recreational use of the dunes, so the period prior to 1970 was marked by intense recreational use. In the late 1960s and early 1970s, ownership of the site passed to a single owner. Use was rationalized which led to the intensification of use in some areas and reduction elsewhere. In 1973, Clwyd County Council adopted a policy of dune conservation, following which there has been considerable improvement in the quality of the dunes. The intractable area of degraded dunes is a consequence of a failure to implement a strategy of car park provision due to a lack of funds.

Other types of policy which affect the environment qualitatively can also be assessed. For example, Wathern *et al.* (1985*b*) were able to describe the impact of the Less Favoured Area (LFA) Directive in parts of upland Clwyd. The demise of semi-natural vegetation in upland Clwyd in the post-war period has been shown to be related to the number of sheep present in the area. Since the introduction of the LFA Directive, increasing sheep numbers have been associated with the availability of LFA aid in the form of headage payments and improvement grants. Over this period, mean farm size has increased as farmers with large holdings and, therefore, deriving greatest benefit from grants and subsidies, have bought out their smaller neighbours. Thus, the policy has done little to sustain rural populations, one of its stated objectives. Although such changes can be identified and described it is unlikely that a rigorous statistical analysis can be undertaken in all cases.

Conclusions

Policy appraisal has the potential to become a valuable tool for environmental management. It would provide a means for assessing the likely consequences of particular policies as they are being formulated. This might result in some policies which are likely to have significant environmental impacts not being implemented. Secondly, after implementation a review could be undertaken to establish not only that the desired objectives are being realized, but also that there are no untoward effects upon the environment. Thus, there are clear parallels between assessing policy impacts prior to implementation and a project-based EIA. In addition, assessing the impacts of a policy after implementation corresponds to a post-development audit.

The applicability of existing EIA methods to policy appraisal has not been established during this research, although it seems likely that the more diffuse nature of many of the impacts of a policy and the wide geographic area that might be affected may place practical constraints upon their use. An important consideration to emerge, however, is that the complexity of the information required for policy appraisal could be reduced substantially by careful selection of monitoring sites. Although this would not provide an overall assessment of a policy, it would allow its impacts to be detected at an early stage if the sites were particularly sensitive to the projected change.

The research has also shown the value of a phased approach to policy appraisal. The resources required for a substantive review of the impacts of a policy are such that this should only be attempted if there is a likelihood that the policy has been implemented in such a way that material changes in environmental quality are likely to occur. Thus, there is a need for prior consideration of both the legislation and administrative procedures adopted to implement a specific policy. This is particularly important where implementation may be delegated to an organization which was not responsible for formulating the policy and, indeed, which may not be sympathetic to its objectives. Consideration of implementation of the EEC Shellfish and LFA Directives in Wales has underlined the importance of legislative and procedural review.

Analysis of a number of policies related to water quality, land use and air quality indicate that the impacts of policies can be detected. It is a complex phased process. Initially, changes in the environment must be detected. Subsequently, these changes must be ascribed to causative actions, which in turn must be shown to be a consequence of the policy under review.

Analysis of environmental trend data to determine the impact of a particular policy probably can never be reduced to a purely statistical exercise. Rarely will policy effect instantaneous change, and frequently it may only modify the rates of change in environmental parameters. The primary objective of statistical analysis should be to determine the timing of such changes within the environment. Certain techniques such as cumulative sum analysis, may help to resolve this difficulty. These calculations, however, will provide only pointers to the possible causes of change. Careful, and sometimes intuitive, analysis will often be required not only to link environmental change to a particular policy, but also to unravel policy mediated change from modifications caused by other factors.

REFERENCES

Bisset, R. (1984) Methods for assessing direct impacts, in Clark, B.D., Gilad, A., Bisset, R. and Tomlinson. P. (eds) *Perspectives on Environmental Impact Assessment*. Dordecht: Rheidel, pp. 195–212.

Brown, I.W., Wathern, P., Roberts, D.A. and Young S.N. (1985) Monitoring sand dune erosion on the Clwyd coast, North Wales. *Landscape Research*, 10 (3), pp. 14–17.

Clark, B.D., Bisset, R. and Wathern, P. (1980) *Environmental Impact Assessment: A Bibliography with Abstracts*. London: Mansell.

Clark, B.D., Bisset, R. and Wathern, P. (1981) The British Experience, in O'Riordan, T. and Sewell, W.R.D. (eds.) *Project Appraisal and Policy Review*. Chichester: Wiley, pp. 125–53.

Council of the European Communities (1975) On mountain and hill farming in certain less favoured areas (75/268/EEC). *Official Journal*, L128, 19.5.75, pp. 231–66.

Council of the European Communities (1976) Covering the quality of bathing water (76/160/EEC). *Official Journal*, L31 5.2.76, pp. 1–7.

Council of the European Communities (1976) On pollution caused by certain dangerous substances discharged to the aquatic environment of the Community (76/464/EEC). *Official Journal*, L129 18.5.76, pp. 23–29.

Council of the European Communities (1978) On waste from the titanium dioxide industry (78/176/EEC). *Official Journal*, L54, 25.2.78, pp. 19–24.

Council of the European Communities (1978) On the quality of freshwaters needing protection in order to support fish life (78/659/EEC). *Official Journal*, L222, 14.8.78, pp. 1–10.

Council of the European Communities (1979) On the quality required of shellfish waters (79/923/EEC). *Official Journal*, L281, 10.11.79, pp. 47–52.

Council of the European Communities (1980) On air quality limit values and guide values for sulphur dioxide and suspended particulates (80/779/EEC). *Official Journal*, L229, 30.8.80, pp. 30–48.

Council of the European Communities (1983) Third programme of action on the environment (1982–86). *Official Journal*, C46, 17.2.83, pp. 1–16.

Countryside Commission (1984) *Evidence to the House of Commons Sub-committee on Agriculture and Environment*. Cheltenham: Countryside Commission.

De Jongh, P. (1982) *EIA and Plan-making in the Netherlands*. Paper to the WHO training course on EIA, Aberdeen, 4–17 July 1982.

Ellis, J.C. and Lacey, R.F. (1980) Sampling: defining the task and planning the scheme. *Water Pollution Control*, 79, pp. 452–67.

Environment Canada (1974) *An Environmental Assessment of Nanaimo Port Alternatives*. Ottawa: Environment Canada.

Environmental Protection Agency (1976) Ocean dumping, proposed revisions to regulations and criteria. *Federal Register*, **40**, pp. 220–9.

Environmental Resources Ltd (1981) *Environmental Impact Assessment: Studies on Methodologies, Scoping and Guidelines*. London: ERL.

Haigh, N. (1984) *EEC Environmental Policy & Britain*. London: Environmental Data Services Ltd.

Hall, R.C. (1977) MEIRS – a method for evaluating the environmental impacts of general plans. *Water, Air and Soil Pollution*, **7**, pp. 251–60.

Hopke, P.K., Gladney, E.S., Gordon, G.E., Zoller, W.H. and Jones, A.G. (1976) The use of multivariate analysis to identify sources of selected elements in the Boston urban aerosol. *Atmos.Environment*, **10**, pp. 1015–25

In't Anker, M.C. and Burggraaff, M. (1979) Environmental aspects in physical planning. *Planning and Development in the Netherlands*, **11**, pp. 128–45.

Jost, D. and Sartorius, R., (1979) Improved air quality due to lead in petrol regulation. *Atmos.Environment*, **13**, pp. 1463–5.

Lee, N. (1983) Environmental impact assessment: a review. *Applied Geography*, **3**, pp. 5–27.

Lee, N. and Wood, C. (1980) *Methods of Environmental Impact Assessment for Use in Project Appraisal and Physical Planning*. Department of Town and Country Planning, Paper No. 7, Manchester University, Manchester.

Leopold, L., Clarke, F.E., Hanshaw, B.B. and Balsley, J.R. (1971) *A Procedure for Evaluating Environmental Impact*. US Geological Survey Circular 645, Washington, DC: US Geological Survey.

MacEwen, M. and Sinclair, G. (1983) *New Life for the Hills*. London: Council for National Parks.

O'Riordan, T. and Derrick Sewell, W.R. (eds.) (1981) *Project Appraisal and Policy Review*. Chichester: Wiley.

Roy, R. and Pellerin, J. (1982) On long-term air quality trends and intervention analysis. *Atmos.Environment*, **16**, pp. 161–69.

Sorensen, J.C. (1971) *A Framework for Identification and Control of Resource Degradation and Conflict in Multiple Use of the Coastal Zone*. Masters thesis, Department of Landscape Architecture, University of California, Berkeley.

Thor, E.C., Elsner, G.H., Travis, M.R. and O'Loughlin, K.M. (1978) Forest environmental impact analysis. *Journal of Forestry*, pp. 723–5.

US Department of Housing and Urban Development (1977) *Integration of Environmental Considerations in the Comprehensive Planning and Management Process*. Washington, DC: US Department of Housing and Urban Development.

Warren Spring Laboratory (1982) *The Investigation of Air Pollution. Directory Part I: Daily Observations of Smoke and Sulphur Dioxide*. Stevenage: Department of Industry.

Wathern, P. (1984a) Ecological modelling in impact analysis, in Roberts, M.R. and Roberts, R.D. (eds.) *Planning and Ecology* London: Chapman and Hall, pp. 80–98.

Wathern, P. (1984b) Methods for assessing indirect impacts, in Clark, B.D., Gilad, A., Bisset, R. and Tomlinson, P. (eds) *Perspectives on Environmental Impact Assessment*. Dordecht: Rheidel, pp. 213–31.

Wathern, P., Brown, I.W., Roberts, D.A. and Young, S.N. (1983a) *An Approach to Policy Appraisal*. Report to the Directorate General for the Environment, Consumer Protection and Nuclear Safety, Commission of the European Communities.

Wathern, P., Brown, I.W., Roberts, D.A. and Young, S.N. (1983b) *Some Effects of EEC Policy*. Report to the Directorate General for the Environment, Consumer Protection and Nuclear Safety, Commission of the European Communities.

Wathern, P., Brown, I.W., Roberts, D.A. and Young, S.N. (1985a) *Policy Appraisal: A Summary and Analysis of Progress*. Report to the Directorate General for the Environment, Consumer Protection and Nuclear Safety, Commission of the European Communities.

Wathern, P., Brown, I.W., Roberts, D.A. and Young, S.N. (1985b) *Monitoring Policy Impact in the Uplands and Coastal Zone*. Report to the Directorate General for the Environment, Consumer Protection and Nuclear Safety, Commission of the European Communities.

Wood, C. (1982) *EIA in Policy Formulation and Planning*. Paper to the WHO training course on EIA, Aberdeen, 4-17 July 1982.

Woodward, R.H. & Goldsmith, P.L. (1964) *Cumulative Sum Techniques*. ICI Monograph No. 3. Edinburgh: Oliver & Boyd.

Young, S.N. and Wathern, P. (1984) The EEC shellfish directive in Wales. *Water Science and Technology*, **17**, pp. 1199-209.

ACKNOWLEDGEMENT

This chapter is based upon research undertaken for the Commission of the European Communities, entitled 'The evaluation of methods for assessing the environmental impact of Community policies – a study of Wales'. The work was conducted under the aegis of the Prince of Wales Committee. This chapter does not necessarily reflect the views of either the Prince of Wales Committee or the Commission with respect to the research, nor necessarily anticipate the future position of the Commission with respect to policy appraisal.

Chapter Seven

EIA and the Assessment of British Planning

MICHAEL CLARK

Even if Britain had not adopted a formal system of mandatory environmental impact statements, the European Commission proposals for Community-wide EIA would have been of value in asking awkward questions about the effectiveness and scope of existing planning procedures. Critical investigation is needed into the relationship between general planning objectives, underlying policies and the often arbitrarily compartmentalized, unreasonably self-regulating systems of development control, infrastructure investment and management, and public sector social and economic intervention which provide the institutional framework for town and country planning in the United Kingdom. EIA has been a step in this process even if its eventual implementation may appear to shore up the *status quo*.

So far most attention has been given to the procedural and organizational implications of EIA: its potential and disadvantages as an addition to the British planning system. Favourable comments frequently contain criticism of this system's effectiveness or fairness, and generally assume that it is a good thing for the environment to be treated in a more explicit and positive way. Hostility tends to concentrate on the administration cost and uncertain benefits of EIA, though there is an underlying tendency to disagree with its sometimes implicit regard for environmental values.

New Planning Control on Farmland and Forestry?

Environmental impact assessment's capacity to extend the scope and comprehensiveness of planning controls has probably been a less obvious feature of recent EEC initiatives. Work such as Marion Shoard's *The Theft of the Countryside* has become widely respected and presents a dismal catalogue of official failure to cope with the environmental and landscape implications of commercial farming and forestry operations (Shoard, 1980). However, the idea of simply extending United Kingdom development control procedures to cover all forms of land-use change has, until recently, appeared so revolutionary as to be outside 'mainstream' political or professional thinking (Labour Party, 1983, 1986; Holbeche, 1984; Ecology Party, 1983).

One of the main reasons for the United Kingdom's recent hostility to the Commission's EIA proposals has been fear that effective environmental or landscape controls should be imposed on Britain's farming and forestry industries. An example of Conservative hostility was seen at the European Parliament's February 1982 debate on the Commission's 23rd Draft Directive on Impact Assessment. Here James Provan, Conservative MEP for North-East Scotland, went against the generally favourable tenor of debate to argue for a flexible and pragmatic approach to environmental impact assessment and for simplification rather than additional 'blanket legislation' and delay:

> to take the major step of moving from no controls on agriculture and forestry, as far as environmental assessment is concerned, to placing these industries in Annex I of the Commission's proposals, along with nuclear power stations, petrochemical complexes and blue asbestos manufacturing, is totally ridiculous.

While this statement is useful in admitting the limitations of Britain's rural planning machinery, and in emphasizing the degree of change proposed, Provan's argument was based on little more than a personal assertion of faith:

> History shows that agriculture and forestry are the proved custodians of the countryside. If agriculture prospers, so does the rural environment, if agriculture is squeezed, another layer of planning bureaucrats is going to make the matter even worse. Present arrangements, including planning permission requirements, are working satisfactorily. (European Parliament, *Official Journal*, 18.2.82, No.1–280/245)

To the next speaker these were merely facile statements:

> Mr President, how Mr Provan, who normally speaks quite sensibly, can say that if agriculture prospers so does the rural environment, is beyond me. I challenge him to back this up with hard facts and not just make facile statements.

There is no obvious way of gauging the influence of Provan's complacent assessment of the present lack of control on agriculture, or the political power of various farming, forestry and landowning interests who stand to lose from additional administrative constraints on their use of land. However, it does seem paradoxical, at a time when there was much academic and professional concern at the inadequacies of Britain's rural planning system, that an EC initiative which offered a basis for comprehensive and even-handed environmental protection should be met by little more than conservative objection. The relative obscurity of successive draft Directives, over-selective processes of consultation and conservationists' obsessive interest in particular threats to the environment (individual big schemes such as nuclear power stations, airports, motorways, moorland enclosure and ploughing, etc.) all help explain why the conservation lobby and ecology movement failed to identify and give effective support to aspects of the Commission's proposals that might have helped counter some of the faults of the 1981 Wildlife and Countryside Act. Subsequent events suggest that it would be wrong to make too much of the slow progress of EIA within the European Community. EIA has implications for the comprehensive and public investigation of any change or proposal likely to do significant environmental damage, and is likely to create a political climate in which it will be difficult for some forms of effective control or public intervention not to follow a negative assessment.

The same climate will impose political penalties on administrations which use arbitrary criteria (such as inclusion in Annex II) to exclude from assessment anything which has the obvious potential for serious environmental impact. Such likely effects of EIA do not necessarily mean that the procedure has brought about a change in values. In part, EIA reflects the general growth of support for environmental priorities, and it is also likely that its implementation in the United Kingdom will be preceded by other important initiatives associated with both the general 'greening' of British politics in the mid-1980s, and with the increased influence of explicitly green pressure groups and politicians.

Military Impact Assessment?

Although any form of Community impact assessment procedure will clearly be structured in terms of categories of land-use change or development, and will be conditional on essentially political decisions about the 'significance' of various forms of impact, implementation of EIA will make it obvious if certain categories of activity are excluded. Farming and forestry are the most obvious, and large-scale, current exceptions but others are also important. At a time when several British local authorities have declared themselves 'nuclear free zones', and when defence has become a topic of controversy within the planning profession, it is worth mentioning that one of the more insignificant amendments to the 1982 European Parliament debate on EIA concerned the inclusion of military installations (European Parliament, *Official Journal*, 18.2.82, No.1–280/242, Mrs Squarcialupi). Commission reaction to this amendment was that 'the time is not yet ripe for dealing with this subject in this directive' (*ibid*, No. 1–280/248, Mr Narjes). This is a pity as EIA seems admirably suited to the regulation and environmental scrutiny of what otherwise appear to be the activities and investment programmes of autonomous military organizations.

United Kingdom Ministry of Defence (MoD) use of a 147-page environmental impact assessment (Ministry of Defence and Property Services Agency, 1984) to support its 1984 plans for the expansion of submarine facilities at Faslane and Coulport, on the Clyde, may be interpreted as a public relations exercise.

However, the case is rather more complex and has wider implications. Proposals in the early 1980s for significant expansion of naval facilities to serve the new Trident weapons system led to a number of objections, to calls for a public inquiry and for the setting up of a Planning Inquiry Commission, and to an EIA exercise and subsequent non-statutory inquiry (Reid, 1983). Both were organized by Strathclyde Regional Council, one of the elected local planning authorities for the area. Although the scheme is 'Crown' development, and so has deemed permission, the MoD sought agreement with Dumbarton District Council, the elected body responsible for development control and local planning matters (Reid, 1983, p. 4).

When agreement could not be obtained within the period of two months specified in SDD Circular 49/1977 (Essery, 1985), the scheme, together with its supporting environmental impact assessment (Ministry of Defence and Property Services Agency, 1984), was referred by the MoD to the Secretary of State for Scotland who has authority over land-use planning and environmental matters in this part of the United Kingdom. In August 1984 he decided to handle the case on the basis of written representations and as a local planning matter (Milne, 1984). Interested parties were invited to submit relevant comment, but there was no public inquiry.

Issues such as the need for this particular strategy for nuclear defence, its local implications in the event of war, and the risk of explosion or radiation discharge in peacetime accidents, or due to terrorist attack, appear to have been excluded by the Secretary of State's terms of reference. In contrast, non-defence matters were given close attention. Approval in March 1985, was conditional on a major landscaping exercise and a number of amenity safeguards and traffic restrictions. The underlying dispute about the need for the base remains and it is clear that no procedure which facilitates the renewal and expansion of NATO's nuclear capability will be acceptable to those who are campaigning for a non-nuclear defence policy, or who are hostile to the NATO alliance. In addition, some who support the strategy of nuclear deterrence will not welcome its facilities in their neighbourhood, though such (genuinely local) objection is likely to be offset or overwhelmed by business and employment interests which are threatened by any reduction in the level of local defence spending. In these circumstances it can be argued that MoD acted sensibly in facilitating criticism of the details of its proposals, while avoiding discussion on their wider context or implication. As it already had authority to go ahead without public consultation, its dealings with local councils and public agencies responsible for environmental and planning matters may be seen as cosmetic: a fiction of legitimization in terms of local impacts. On the other hand, the resulting modifications met many of the minor, conventional planning objections to the initial scheme and put the onus of non-cooperation on those, including the statutory local planning authorities for the area, who continued to object to the essential nature and underlying purpose of this nuclear base.

Without consensus, the effect has been to reinforce each side of the argument.

In some way the frustrations of local politicians and pressure groups opposed to the Trident base contributed to the groundswell of anti-nuclear sentiment which led, by 1986, to the most serious consideration yet of non-nuclear defence options by the Labour and Liberal parties. It may be too strong to suggest that sham versions of local consultation undermine their fundamental objectives, but in this case the process of legitimization has clearly been unacceptable to many who would, in more normal circumstances, participate in the decision-making process.

Legitimization

A common system for civil and military impact assessment would reassure some of those who doubt the internal planning competence or environmental awareness and sympathy of decision makers within the services and associated Civil Service. However, it is doubtful if security considerations could ever permit impact assessment of military projects to be as open, or as wide-ranging in its consideration of alternatives, as that for civilian schemes. In addition, military decisions are central to government authority within each nation state. Political responsibility, scrutiny and control are jealously maintained at the highest levels of administrative and executive decision making. In these circumstances it is difficult to see how the public participation, consultation or local democratic involvement elements of impact assessment will ever be permitted to be more than an enabling form of public relations exercise.

What is significant in recognizing the real political limitations of applying what would amount to a form of local legitimization to isolated aspects of national or supra-national defence policy is that there are strong parallels between defence and many other aspects of central government policy. In Britain town and country planning enjoys a large measure of apparent devolution. Elected local councils play a key role in overseeing their professional staff's plan-making and development control functions. Ministerial authority is generally remote and expressed through enabling legislation, plan approval, appeals against local decisions and general policy guidelines.

Unfortunately the effectiveness of local and interest group activity in delaying or thwarting unwelcome change, and in obliging would-be developers to anticipate and circumvent objection, has given the false impression of a requirement that projects be assessed in terms of the people they directly affect. While disadvantageous change clearly deserves compensation, assessment on the basis of interest group and community reaction is likely to be regressive. Interest and community group membership often reflects above average income or lifestyle and it can be argued that in protecting their personal interests such groups harm those of much larger, but less well off and less well represented, sectors of society. Less appropriate site selection, higher operating costs or denial of potential benefit to large numbers of people may result (Eversley,

1976, p. 131; Kaufman, 1973). On the other hand there is the powerful argument that organized objection to big schemes is necessary to identify faults in design and site selection, and to give some political support to social, aesthetic, environmental or cultural values which would otherwise be submerged (in the national interest).

One can interpret Britain's system of public inquiries into large projects from various standpoints. Sympathy with the tasks and responsibility of central authority will emphasize the generous way in which minority interests and local groups are allowed to have a say in important matters, partly for the pragmatic reason that their contributions are of value to ministerial decision makers. This 'centralist' position tends to favour procedural improvements which increase the scientific or objective component of site specific inquiries and consultation. EIA has potential here, like cost benefit analysis, as a set of clearly laid down administrative steps which 'grease the tracks with rationality in order to avoid emotional impediments to technological advance' (Green, 1976, p. 942).

The possibility of allaying local fears and pre-empting the growth of effective protest, with its resulting delay and uncertainty, seems to explain most environmental impact statement preparation in the United Kingdom (Petts and Hills, 1982). These voluntary adjuncts to public inquiries and ministerial decision making have also been recognized to improve the quality of site choice and programme planning, though in the case of British Gas this favourable discovery has been used to support the corporation's view that existing procedures are satisfactory, and that there is no need for mandatory EIA (Dean, 1981).

Limitations of the United Kingdom's Public Inquiry System

Criticism of Britain's established system of public inquiries into major projects can be based on the simple argument that the system no longer works. It fails to provide politically acceptable decisions within a reasonable timescale, is unjust in its allocation of advantages and costs and operates on a different, to some extent, incompatible, set of assumptions about the location of political power from its formal position as a source of advice for executive ministerial authority. Confusion about its (officially non-existent) 'local legitimization' function is enhanced by inquiries' quasi legal status, by the extravagant employment of members of the legal profession and by the system's damaging tendency to inhibit ministerial publication of strategic guidelines or general policy statements. Even participation in academic discussion is affected:

> Whilst the department keeps abreast of development in the fields you mention, it has to remain detached in case of statutory involvement at a later stage.
> (Departments of Environment and Transport, 1982)

Something seems to be wrong here. While the 'pluralists', those who believe in the potential and inherent justice of local processes of legitimization, seek to enhance and regularize their position, central authority finds itself blocked by both the political results of their action, and by the constraints it imposes on itself to protect its statutory position as final arbiter. In a search for a way out of Britain's delay ridden and costly public inquiry system, EIA has been grasped by pluralists as well as centralists. Its assumptions about freedom of information, prior publication of evidence, consideration of alternative sites and actions, and the need to justify environmental damage, however caused, are all welcome to people who believe that their involvement is helping to increase the democratic accountability of government, as well as defending interests which they personally value.

Policy Issues

The search for improvement in public inquiry performance has also been directed at the general context for major planning decisions. Although important policy questions continue to be aired at site-specific inquiries, statutory provision remains for a Planning Inquiry Commission to be held into a proposal 'where there are considerations of national or regional importance' (MacDonald, 1982). Successive governments have avoided this possibility, to the extent of holding lengthy inquiries at Windscale, Sizewell, Stansted and Dounreay at which the main issues in contention had little to do with local planning. What was in question was national policy – in these cases for energy and airport planning. However, the structure of the inquiry here and in similar cases has permitted the Minister to appear aloof. It has been up to the industries or public institutions proposing development to establish need and relate this to specific sets of proposals, while opponents have been able to question the necessity of development and the suitability of the proposed design. The result has been to minimize effective criticism or review of national policies, and to encourage government departments either to do without formal policies or strategies in major areas of responsibility, or to conceal the fact that such policies exist.

The quasi-legal nature of public inquiries into major development proposals, and the short-term political advantages of flexibility, may explain why clear policy guidance and properly researched national planning strategies are largely absent from the British planning system, at least as far as it relates to major investment projects of the type that result in hotly contested public inquiries and difficult, often delayed, ministerial decisions. A significant exception is the influential and comprehensive Scottish Development Department series of *National Planning Guidelines*. The problems caused by such a policy vacuum have led to considerable support for some form of two stage inquiry system, with assessment of the merits and disadvantages of detailed proposals for a specific site coming after a more general inquiry into the need for this type of investment. The idea seems similar to that of a planning inquiry commission, but would probably involve Parliamentary debate as well as expert scrutiny and adversarial conflict (Council for Science and Society . . ., 1979; Shore, 1978). Although this may be seen as a centralist plot to establish need for investment without reference to the damage new schemes will cause, most enthusiasts for two-stage inquiries see them as a way of applying the values and procedures associated with existing inquiries at a more appropriate scale (Le Las, 1982; Clark, 1980).

As one approaches matters of national importance, and as the element of policy criticism increases and the range of future options is narrowed, so the parallels between civil and defence planning become obvious. No national administration will allow key areas of policy making to be determined by factional or parochial interest groups. It is naive to suggest that by merely extending the structure of public inquiries, or the values of EIA, one can force an unwilling or cautious government to embrace the political difficulties of rigorous and public policy appraisal, particularly when this is to be dominated by a 'higher' (and unquestioned) set of environmentally based values.

If policy appraisal is to be attempted under similar quasi-legal arrangements to existing public inquiries then the effects on central government's policy guidance and review activities could be most counter-productive. There is a considerable risk that ministerial detachment, and supposed objectivity, will be further protected by avoidance of topics that might be the subject of future inquiries.

The Need for Ministerial Guidance

What is needed is a broad framework of national plans and policies for matters best dealt with at this scale: energy policy, communications, regional development etc. Obviously such plans and policies must be based on the most thorough research, be regularly updated or changed to meet new circumstances, and must be flexible or general enough to accommodate commercial requirements and local considerations. Legitimization at a national scale is necessary. This will involve Parliamentary debate and an element of Cabinet commitment, though future governments may also consider it appropriate to carry out public participation exercises such as the recent referenda on nuclear power in Sweden and Denmark and the national debate on energy policy in West Germany.

Such participation would be a break with the United Kingdom tradition, but it may be necessary if we are to escape an *ad hoc*, unrepresentative, unjust and legally constrained system of policy making, or avoidance, associated with today's big inquiries. It can be argued that the political contention associated with large projects, particularly where they involve the nuclear industry, has made them impossible for the development control system to handle. As many of the issues are common to those involved in the implementation of defence policy it is rational that decisions should be made nationally, rather than in an artificially local context.

A move towards centralized policy making may be seen as the withdrawal of democratic rights and of environmental and community safeguards associated with local public inquiries. Hopefully, the opposite will occur. Ministerial policy will be based on up-to-date and public information, and will be subject to the constant attention of Parliamentary scrutiny as well as the continued attention of interest and environmentalist groups. The fiction of Ministerial impartiality, through quasi-judicial status, will be replaced by obvious responsibility. Local inquiries will still be necessary to cope with local matters and genuinely inform decision makers, and will be well served by the Community's mandatory environmental impact statements. An EC Environment Programme based on uniform standards of pollution prevention and reduction may also help by providing rigorous legal controls to augment or replace Britain's somewhat permissive approach to the control of industrial waste and emissions (Commission of the European Communities, 1979; Clark, Bisset and Wathern, 1981; Tyrell, 1981; Haigh, 1984).

Here, as in the extension of United Kingdom Planning Controls to cover agriculture, it can be argued that a learning process is underway. The House of Commons Select Committee on the Environment which dealt with the planning inquiry system in 1986 saw a need to separate national policy and site specific issues:

(i) Wherever possible national policy should be clearly determined and laid down by government prior to a planning inquiry on a major development which should take place within the context of such policy;

(ii) As far as is practicable, the questioning of national policy determined elsewhere should not be allowed to impinge upon public inquiries, which should be concerned with site specific issues and with whether a particular site fits within the national policy;

(iii) Where national policy considerations are unavoidable in respect of a major development and government needs a public inquiry to determine that policy, it should:
(a) in a development proposal which involves a number of alternative or successive sites, seek the assistance of a Planning Inquiry Commission, for which statutory provision exists, before using site specific public inquiries;
(b) in remaining cases, use a unitary major public inquiry modified by the use of the DoE's new Code of Practice.

The RTPI's 'Agenda for the next Parliament', presented in November 1986, called for important development proposals to be given national debate before being authorized. Multi-stage public inquiries should be introduced, and be based on national and regional policy guidelines for matters like energy and transport.

The Committee of Inquiry chaired by Lord Flowers, which reported to the Nuffield Foundation in 1986, made similar recommendations:

Central government should publish from time to time concise and consistent statements of national policy wherever national interests in land use and development are at stake, and this should be its primary planning function. (Flowers, 1986, p. 186)

Such priority stemmed from a general frustration at the lack of coordination between land-use planning and other forms of planning, and also between different arms of government (*ibid*, p. 68). There was also 'widespread criticism' at the lack of national policy guidance (*ibid*, p. 72). Here the TCPA submission to the Nuffield Foundation's inquiry illustrates the case against centralism, but for strategic guidance:

> Many functions should in fact be devolved from the national level to a regional level of government in relation to virtually all major items of infrastructure, strategic decision-making and resource allocation. Nevertheless, what is required at the national level that is not at present attempted is a continuous public discussion about national policies in the major key areas of energy, communications, population and employment distribution. There should be a proper fora for informed discussion of these issues with a view to the framing of appropriate policies. (Hall, 1983, p. 10).

It remains to be seen if this recent political and professional enthusiasm will overcome bureaucratic caution, and result in a change in the way the United Kingdom planning system copes with large-scale, difficult or general issues. Academic and practical arguments for change appear to have been influential, in particular Jennifer Armstrong's paper on the Sizewell Inquiry and evidence to the Select Committee (Armstrong, 1985; Environment Committee, 1986, p. lxv). A shared perspective and set of values has probably also led a number of individuals to reach similar conclusions to those outlined in this chapter: a function, in part, of the intellectual environment. More rigorous motivation can be seen in the context of a major government initiative to do without prior consultation with planning authorities or public inquiries for large schemes (Flowers, 1985, p. 1).

Possibly because of unforeseen use of the relaxed planning controls in an Enterprise Zone, the Canary Wharf project on the Isle of Dogs has escaped at least the initial rigours of ministerial scrutiny and local hearings. Major decisions about a Channel Tunnel have been based on quickly produced EIA's and appear to have been influenced more by the public relations gimmickry of rival schemes than by considered attention to local or national implications. There has been criticism of use of Parliamentary hearings as a substitute for a public inquiry, but the system has provided a convenient platform for rival opinions, and has allowed the 'facts' of the case to be explored openly, efficiently and within a few months.

As with more ponderous public enquiries, criticism of the Channel Tunnel hearings has included objection to the scheme's policy context and too early commitment to a specific design (Levin, 1976). There is also resentment at the legitimization element of investigation of a scheme which the government already openly supports and may even have initiated. Here a committee which reflects the balance of power in Parliament is perhaps a more honest mechanism than a supposedly independent Inspector, though it may be argued that a proposal would have to be seriously flawed before either approach could deflect the political and bureaucratic inertia behind a favoured project.

Implications for Public Corporations: Restore Co-ordination and Accountability

Once national authority has been restored in the fields of energy and communications planning it will be appropriate to look at decision making and accountability within the relevant industries. The same need for assessment follows from consideration of the way that large corporations, and associated public bodies, support their investment programmes at public inquiries. The electricity, nuclear fuel, coal and gas industries have been lavish in using public funds to support major development applications which follow logically from established practice and conform to current thinking within the responsible professions (Clark, 1983). However, many of the difficulties which such applications face can be accounted for by lack of public review of the underlying policies. Local planning inquiries have become a rare opportunity to question policy, though primarily in terms of a project's suitability in a specific site. This permits the corporation or agency proposing development to make assumptions about public policy corresponding with industrial practice. In particular the narrowly defined operating goals and statutory duties of public corporations become an excuse for excluding options which conflict with an organization's overriding purpose. Core objectives which are enshrined in statute acquire an inertia and permanence which itself seems immune from policy review.

The fossilised goals and overriding operational (and operator) interests which result contribute much to the difficulty facing ministerial decision makers. The main initiative on public sector investment comes from the industries themselves. While there are advantages for central government in appearing detached from the acrimonious conflict which generally ensues between a public corporation and various objecting interest groups, it is difficult for legitimate objection to be respected without putting a whole scheme in jeopardy.

Much of the conflict at recent contentious public inquiries has been between public bodies' interpretation of key objectives and viable options, and the conflicting interpretations of goals and possibilities made by objectors' expert witnesses. Organizations such as the Central Electricity Generating Board (CECB), Civil Airports Authority, British Nuclear Fuels Limited and National Coal Board (now 'British Coal') have extended their commercial and initial public utility functions to become major contributors to the future pattern of land use, economic opportunity and environmental quality within the United Kingdom. British Rail, British Gas, British Telecom, British Airways, the National Health Service and the regional water authorities can be added to the list of public or recently 'privatized' bodies whose corporate plans, together, amount to a massive commitment of capital spending. Amazingly most of these public or publicly-regulated funds are allocated in isolation from any form of national land-use or regional development planning strategy. Corporate planning is a strictly internal business which, although in many cases subject to ministerial review and responsive to day to day political intervention, has not been formally integrated with the 'Planning' function of government.

Replace Autonomous Public Corporations with Direct Control by Regional Government

If policy guidance and review in the fields of land use, economic management and environmental planning can become an area of clear ministerial responsibility and intervention, rather than one of cautious statutory involvement as final arbiter, then there is considerable scope for rationalizing and integrating the planning related activities of public agencies. While some bodies may merely perform their statutory functions more effectively if properly integrated with other public organizations, others need to be brought within direct governmental control if the tendency for operational interests to override wider public interests is to be curtailed. Present lack of political accountability suggests that there will be advantages in devising some form of regional level of government to accommodate the existing regional water authorities, and possibly to take over functions now performed at an inappropriate and unaccountably large scale by the CEGB, British Gas, British Rail and British Telecom.

The recent rush of public sector asset stripping and the associated 'privatization' of several large public corporations may have pre-empted such a rational approach to public sector restructuring. However, it may also have prepared the ground for new forms of regulations and policy guidance. 'Planning agreements' and other explicit forms of direction require a clear policy context and may offer an effective alternative to renationalization, though they may also be seen as second best, and an abandonment of socialist principles. It is conceivable that a return to public ownership could do little to overcome the utilities' recent lack of integration or accountability, whatever the rhetoric supporting reorganization.

There are dangers of turning frustration at various organizations' disagreeable policies into schemes for redefining control on terms which reflect one's own interests or values. This is perhaps legitimate where a body such as the CEGB permits itself to become committed to a narrow interpretation of the options open to it, though here rejection of anything other than a programme of large-scale electricity generation, based on nuclear power or coal, seems to reflect ministerial attitudes as much as those within the industry. The extreme (if not luddite) measure of disbanding the CEGB and dispersing its powers to some newly elected system of regional government might involve penalties in lost economies of scale and a hindered nuclear research programme. On the plus side would be far easier integration of schemes involving combined electricity generation and waste disposal (using domestic rubbish as a low grade fuel) or district heating. It should be easier for a locally elected body to find acceptable sites for new power stations, and such a body should be more responsive to local deficiencies in generating capacity, as found in Devon and Cornwall.

At very least the disbandment of our great and autonomous corporations, and their incorporation within a new tier of regional government, might have a minor but aesthetically vital effect on Britain's roads. It would mean that some degree of co-ordination might be possible between the utilities and statutory undertakers responsible for digging up our streets, and the county road authorities which seem incapable of requiring them to be restored to their former state.

Conclusion

While EIA has several positive implications for British planning its main implications are oblique. There is much good in the European Community's recent environmental initiatives. EIA may make Britain's public inquiry system more efficient and just. However, its main potential lies in the implicit extension of planning control to all forms of land-use change. This would be of great benefit to landscape planning and environmental management in farmland and forestry. The idea of applying EIA to policy faces major problems. British land-use planning is heavily constrained by the statutory convention that the Minister should appear detached from the argument presented to him at a public inquiry. As a result there are few policies to assess, even in areas as important as the location and type of power stations, airports or motorways. The local inquiry system goes some way towards filling this vacuum, but at great cost to most of its participants. Many commentators see an answer in a two-stage inquiry system, but there are dangers of confusing local processes of legitimization which have grown up, unofficially, around big public inquiries with a larger element of pluralism than is acceptable to Britain's centralized system of governmental responsibility. The parallels between policy making for defence, and for major civil investment programmes, support the argument that policy review is a function of central government and Parliament.

First, however, there must be policies to review. If Britain's lack of strategic planning guidelines at a national scale is a result of statutory conventions about a Minister's quasi-judicial role, then it is necessary to abolish these conventions. Arguably they are little more than a convenient legal accident which permits executive authority to appear remote from its own decision-making processes, and which allows maximum flexibility. Properly researched, debated and reviewed, national planning strategies would permit a far less cautious, delaying and uncertain process of ministerial decision making than the present reaction to massive public inquiries.

It would also bring the initiative for promoting and guiding investment back to central government, and away from large corporations. Many of these are public agencies which are free to operate with little interference on the basis of what are often narrow and outdated statutory goals and responsibilities. Their scale of uncoordinated investment and operational decision making suggests that the main benefit of clear national planning strategies will be to make these diverse aspects of the public sector contribute positively to wider objectives than those of operational interest and convenience.

The overall impression is that the simple act of trying to make Britain's planning system capable of accommodating impact assessment will lead to a variety of valuable improvements. Just obliging ministries to formalize and publish policy guidelines would have massive implications for the coordinated and positive management of a wide range of public sector institutions. The fact that this would include explicit treatment of environmental issues is almost irrelevant when compared with the other implications of effective policy leadership. EIA can clearly help local planning inquiries, particularly if it introduces an element of open and sympathetic government which looks kindly on participants. However, its main implication is for the extension of effective planning controls into the British countryside. Hopefully the resulting hostility of landed interests will not be successful in its attempt to block or emasculate this part of the European Community proposals.

REFERENCES

Armstrong, J (1985) *Sizewell Report: a New Approach for Major Public Inquiries*, London: TCPA.

Clark, B D., Bisset R, and Wathern, P. (1981) The British experience, in O'Riordan, T. and Sewell, W.R. Derrick (eds) *Project Appraisal and Policy Review*, Chichester: Wiley, chapter 6.

Clark, M. (1983) Taking power in the regions. *Town and Country Planning* **52**, pp. 338–40.

Clark, M. (1980) *Planning Processes and Ports: British Land Use Implications of Maritime Change in the Nineteen Seventies.* Unpublished PhD thesis. UWIST (University of Wales), chapter 9.

Commission of the European Communities (1979) *State of the Environment Second Report.* Brussels: Commission of the European Communities.

Council for Science and Society, Justice and the Outer Circle Policy Unit (1979) The Big Public Inquiry. *A proposed new procedure for the impartial investigation of projects with major national implications.*

Curtis, F. (1982) A checklist for the writing and presentation of environmental assessment (EA) reports. *Canadian Geographer*, **26** (1), pp. 64–70.

Dean, F. (1981) The use of EIA by the British gas industry, in Breakall, M. and Glasson, J. (eds.) *Environmental Impact Assessment: From Theory to Practice*. Oxford: Oxford Polytechnic, Department of Town Planning, chapter 8.

Departments of Environment and Transport (1982) Letter explaining their inability to make a contribution to a meeting organized by the Geography and Planning Study Group of the IBG to discuss the strategic planning implications of current hydrocarbon developments in the Irish Sea.

Ecology Party, *Politics for Life, 1983 Election Manifesto*, p. 15.

Environment Committee (1986) *House of Commons, Fifth Report from the Environment Committee, Session 1985–86, Planning: Appeals, Call in and Major Public Inquiries*, Vol. I, Report together with the Proceedings of the Committee. (House of Commons Paper 181–I). London: HMSO.

Essery, D J. (1985) Notice of proposed development by the Ministry of Defence: Clyde Submarine Base at Faslane and Coulport. Letter to Dumbarton District Council, 7th March. Edinburgh: Scottish Development Department, ref P/SLR/31/2 mimeo.

Eversley, D.F.C. (1976) Some social and economic implications of environmental impact assessment, in O'Riordan, T. and Hey, R. D. (eds) *Environmental Impact Assessment*. Farnborough: Saxon House.

Flowers, Rt Hon the Lord, FRS (1986) *Town and Country Planning: the report of a Committee of Inquiry appointed by the Nuffield Foundation*. London: Nuffield Foundation.

Green (1976) quoted in A B Lovins (1977), Cost-risk benefit assessments in energy policy. *George Washington Law Review*, 45(5), p. 942.

Haigh, N. (1984) *EEC Environmental Policy and Britain*. London: Environmental Data Services Ltd, chapter 1.

Hall, D. (1983) Nuffield Foundation inquiry into the system of Town and Country Planning. Submission by David Hall, Director TCPA. London: TCPA, mimeo.

Holbeche, B. (1984) Planning control and agriculture. *The Planner*, 70(6), p. 20.

Kaufman, A. (1973) Beauty and the Beast: the siting dilemma in New York State. *Energy Policy*, pp. 243–53.

Labour Party (1986) *National Executive Committee. Statements to Conference . . . Blackpool 1986*. London: Labour Party.

Labour Party (1983) Home Policy Committee, *Agriculture and the Environment*, Draft NEC Statement.

Le-Las, W. (1982) The myth of restructuring the Big Inquiry. *The Planner*, 68 (3), p. 36.

Levin, P H. (1976) *Government and the Planning Process*. London: George Allen and Unwin.

Macdonald, K. (1982) The sleeping public inquiry monster. *Planning*, 13.8.82, p. 7.

Milne, R. (1986) Proceedings fuel fast track fear. *Planning* No. 695, pp. 8–9.

Milne, R. (1984) Submarine plans point way forward on impact studies. *Planning*, No. 570, pp. 6–7.

Ministry of Defence and Property Services Agency (1984) *Proposed Development at the Clyde Submarine Base, Environmental Impact Assessment*. London: Merlin Arts Limited.

Ministry of Defence, Defence Lands (1984) Notice of proposed development at the Clyde Submarine base (Faslane and Coulport), 14th May 1984. Mimeo.

Petts, J. and Hills, P. (1982) *Environmental Assessment in the United Kingdom: A Preliminary Guide*. University of Nottingham Institute of Planning Studies.

Reid, J M. (1983) *Report of Inquiry into the proposed extension of the Royal Naval Armaments Depot at Coulport, Dumbarton, for the Trident Weapon System, 20th/24th June 1983*. Glasgow: Strathclyde Regional Council.

Shoard, M. (1980) *The Theft of the Countryside*. London: Temple Smith.

Shore, P. (1978) Secretary of State for the Environment. Statement on Major Planning

Inquiries, reprinted in *Journal of Planning and Environmental Law*, November, 1978, p. 735.

Tyrell, A. (draftsman, 1981) *Opinion of Legal Affairs Committee* 'on the proposal from the Commission . . . to the Council (Doc 1–293/80) for a directive concerning the assessment of the environmental effects of certain private and public projects', Rapporteur Mrs B Weber, European Parliament Working Document 1-569/81/rev, 2.2.82.

Chapter Eight

Environmental Values
in a Changing Planning System

JOHN HERINGTON

Just at a time when the British government is trying to loosen the planning
system – and succeeding to some extent – a certain tightening will thus have taken
place – albeit for a narrow class of development. (Nigel Haigh, 1983)

Optimists argue that the compulsory introduction of EIA procedures into the
British planning system will bring about more informed decisions on large-scale
developments and may create the political climate for bringing under control
some activities which are not now subject to public planning regulation. Much
of the discussion has assumed that EIA *per se* has the potential to change the
planning system in ways which will allow greater expression of environmental
values in public decision making. But this is to ignore the political ideology
necessary to allow this to happen. Very little attention has been given to the
changes underway within the public planning system nor to their influence on
the way in which EIA can be used to have a bearing upon political decisions.

   This chapter starts from the proposition that it is more realistic to view the
emergence of EIA as an outcome rather than a cause of change in the planning
process. Its theme is the changed political context within which planning
decision making is now carried out and the relationship between political and
environmental values. Emphasis is given to three issues: changes in the political
economy and national policy goals; the problems planners face in mediating
between developmental and environmental interests; and the incorporation of
environmental agencies in the decision-making process.

Changed National Perceptions of the Planning System

There is plenty of evidence that the traditional roles played by the public
planning system in Britain are being eroded (Greater London Council, 1985).
Urban containment is giving ground to facilitative modes of planning both at

central and local government levels. Plan formulation and development control processes have been and continue to be subject to substantial modification in favour of private development interests. Significantly, these changes in statutory planning have occurred while local government and private developers have been learning the value of using EIA to solve development problems. In the 1980s favourable government attitudes towards economic development set the tone for planning attitudes towards the environment.

The weakening of the planning control function in local government is associated in part with changed government ideology. The 1979 Conservative administration set out to encourage a market-led economic revival in which the private sector could flourish at the expense of the public sector. Established policies and practices which might inhibit the pursuit of economic growth, including the social welfare system, local government and the trade unions all required reform.

> Planners must help create the right conditions and ensure that business initiatives prosper. What does concern me is that planning procedures should not hamper the economic recovery now slowly emerging . . . Planning authorities must adopt a flexible and pragmatic approach to meet the needs of versatile enterprises. (Rt.Hon Patrick Jenkin, 1984)

The goals of the government's industrial strategy are not evidently in conflict with land-use policy, according to the government. The importance of the environment, countryside conservation and the protection of the Green Belt is reaffirmed by Ministers. Indeed some planning controls have been strengthened, notably in specially designated landscape areas including the National Parks and Areas of Outstanding Natural Beauty. However, the protection and maintainence of environmental quality is seen as a cost on economic production, either in terms of the design standards of new developments or as a cause of planning delays. Although many industrialists now argue that a decent environment is 'good for business' the environmental equation is often presented in narrow accounting terms, without it ever really being demonstrated to what extent the benefits of environmental protection outweigh the costs.

The government's priorities for land planning were revealed in a number of ministerial speeches and the publication of several controversial circulars following the Local Government Planning and Land Act 1980 (Department of Environment, 1981 and 1982). The essential purpose behind these changes was to speed up and simplify the process of producing plans. However, the removal of blockages to private sector development has become an equally important aim of the modifications to the plan-making and control functions of local government. Structure and Local Plans should be flexible enough to accommodate future change. If they are not, substantive modifications may be made by the Secretary of State to overall targets for industrial and housing development, for instance, by increasing the amount of land allocated for new factories. The general effect of these changes in planning policy and practice is to shift the balance in favour of development and against protection and restraint over large tracts of rural Britain outside the major cities (for a review, see Herington, 1984).

The changes in forward planning have taken place within the context of an already permissive development control system. Between 1962 and 1982/3 a very high proportion of planning applications were approved, on average over 83 per cent. Since 1979 the approval rate for major developments was about 75 per cent but fell sharply to about 65 per cent in June 1981 following the introduction of planning charges (Goodchild and Munton, 1985). The percentage rate of approval will barely be affected by the relatively small number of Annex 1 developments which will in future be subject to environmental impact assessment procedures. Morever, even if there were a marginal increase in refusals of projects subject to an EIA, there is little guarantee that refusals by the local planning authority will be upheld by the Secretary of State should the developer decide to appeal.

The number of planning appeals has been increasing fast. In 1985 there were nearly 17,000 and in 1986/87 there were 19,856 recorded. The chances of success on appeal for the owner or developer have (at least for all classes of development) increased since 1971 from 1 in 5 to 1 in 2.5. Appeals determined by the Secretary of State have been markedly more successful than those determined by Inspectors, an important point since these are generally bigger developments. It is precisely these large projects subject to an EIA which are likely to have been 'called-in' and determined by the Secretary of State should there be an appeal.

Absence of Policy Direction

Recent changes in the statutory planning system take place without any clear national direction over planning policy, at least in England and Wales. The Department of Environment is potentially capable of producing an environmental policy for Britain, but this would require coordination between many different sectoral interests in land-use, water, energy, transport, industry and so on. The DoE finds it difficult enough at present to coordinate its own administrative divisions between the directorates of planning, inner cities, housing and construction, finance and local government. The decision to hive off transport to a separate department in 1976 made the integration of transport and land-use planning more difficult. The Countryside Commission, with its broad remit to protect and manage non-built-up areas, was also separated from the DoE in 1982 just at a time when the relaxations in the statutory land-use planning system were being introduced and were likely to have national implications for the future of the countryside. The government's policies on energy, industry and employment are the province of separate departments.

Indeed the relationship between territorial and sectoral policies is notoriously ill-defined in Britain as in other Western European countries (Mckay, 1982). For this reason implementing an agreed set of national environmental priorities seems improbable. The Conservation and Development Programme for the United Kingdom, part of the World Conservation Strategy launched in 1980 favoured greater government priority for resource management, the development of a long-term plan for rural and urban areas, and the greater integration of conservation and economic development programmes. There are formidable obstacles to such an approach. The sectoral policies of central government are influenced by the vested development interests of the transport lobby and construction industry with little reference to the land-based policies of local government. Planning for urban and rural land is functionally and administratively fragmented. At the same time the agricultural and forestry industries are largely exempt from environmental accountability (Johnson, 1983).

It is virtually impossible to obtain a consistent view on the importance of environmental factors in national level decision making. Nor is there in Britain any departmental overview of separate policies, of the kind which the United States operates in which Congress tries to monitor federal activity. Policy review at national level has been proposed by many commentators (see, for instance, O'Riordan, 1981). Similar suggestions have been made about plans and programmes. National level consideration of the principle of public sector programmes has often been proposed as a logical development of EIA procedures. In the case of roads the government has introduced principles by which noise, visual impact and compensation aspects should be judged in relation to the design of new roads, but it has failed to assess the wider environmental policy implications of motorway programmes; for example, as they impinge on urban land requirements and amenity (see for instance, Watkins, 1981).

One of the late modifications to the EC Directive had the effect of exempting matters decided by national government (e.g. Acts of Parliament) from mandatory EIA procedures. The exemption may apply to Parliamentary Bills which can be brought forward by public and private bodies (these Bills effectively circumvent usual town and country planning procedures including public local inquiries, although opponents have a chance to discuss their objections). There are three points to consider: first, whether public sector bodies sponsoring developments which will have a major environmental impact should be permitted the choice of 'one project at a time' approach and, if they are, whether these bodies ought to be first required to demonstrate how their project falls within any overall policy or programme; second, if the petitioners have access to any environmental impact assessment produced under a Private Bill there may be enhanced opportunity for public discussion provided environmental interests have sufficient resources to make an effective petition and that they are well represented in Select Committee; third, and more speculatively, whether more environmentally sensitive decisions will result from more informed parliamentary discussion of Private Bills.

The case of the Empingham Reservoir development Bill (Lawson, 1982) is instructive because it suggests that none of these provisos is automatic. The sponsors, the newly formed Welland and Nene River Authority, obviously thought it would be easier to avoid any general assessment of water industry policy by using the Private Bill procedure. The petitioners, who included the affected local authorities in the area and the Country Landowners Association, Council for the Protection of Rural England and National Farmers' Union, argued that the reservoir was unnecessary and that it would have a detrimental local impact. They were unable to challenge effectively the technical basis of the population forecasts which underpinned the River Authority's case nor to pursue the argument that it would be better to increase existing sources or develop alternative locations. The government's case, presented when the Bill had its second reading, was that the reservoir was a 'regrettable necessity' and that there were no cheap alternative solutions. Although the Bill was debated three times, the petitioners lost their case. The most significant pointer to present times was the extent of disagreement between the House of Lords and the House of Commons and the establishment of a special Select Committee to examine the broader issues of water resource planning at a national scale with the implication that in future the water industry should make a more thorough examination of alternatives before embarking on specific projects.

Most cases of public sector development proposal are handled through the planning procedures and involve, usually (though not always), a public inquiry. Despite the system, the views of the public often do not appear to weigh highly when inquiry decisions are announced. It seems unlikely that greater sensitivity to public feeling would result from the introduction of environmental impact assessment. The reason for saying this is that the prevailing ethos of recession and unemployment and a market-orientated government has encouraged the economic development departments responsible for trade, industry, energy or transport to get their way, sometimes irrespective of the findings of inquiry inspectors. In 1985 several key projects were permitted all within a few months and following major public inquiries (apparently without the relevant departments carrying out any environmental impact assessment). Approval for the route of the M40 motorway was given by the Department of Transport following a major inquiry. The new motorway will be built in open countryside between Oxford and Birmingham and was opposed in principle by the Council for the Protection of Rural England on the grounds that existing road and rail links could be utilized more effectively and would be less intrusive (Council for the Protection of Rural England, 1983). The considerable expansion of Stansted Airport was also announced by the Department of Transport, following a protracted public inquiry and fierce opposition from national and local conservation and planning organizations. The exploitation of oil on a scenically attractive island in Poole Harbour, Dorset, was sanctioned by the Department of Energy without a public inquiry, thus overruling national priorities for the protection of Areas of Outstanding Natural Beauty and flying in the face of opposition from local planning committees.

Underlying these examples is the absence of clear government policy. All such developments would have required an EIA under EC legislation and the need for firm policy guidance would seem to be of paramount importance. The problem is not that government has *no* policy, but that it is very secretive about what its policies are, and that there are no mechanisms for ensuring that policies are based upon adequate research, responsive to key issues (as made explicit by EIA), or safe from factional or sectional bias.

The Weaknesses of the Inquiry System

Public inquiry procedures have come in for much criticism and, more recently, reappraisal of their effectiveness for dealing with major economic development projects. The introduction of formal EIA in 1988 needs to be seen against a background of growing disillusion with current approaches to the public inquiry.

The government has produced a series of proposals for speeding up planning appeals and for making it more costly to appeal without reasonable grounds (HMSO, 1986). Behind these administrative changes is the belief that there should be a presumption in favour of development. Although the balance should be held between conservation and development, improvements in efficient handling of appeals can be made without lessening environmental quality. The government appears to accept that 'its policy advice is not always as accessible to local planning authorities and developers as it could be' and will try to do better in producing clearer policy guidance. On the other hand it rejects the Select Committee recommendation to separate national policy from site specific issues (House of Commons Environment Committee, 1986), an approach widely advocated elsewhere (Armstrong, 1985; Liberal Party, 1986). Their reluctance to define national policy seems to rest on the grounds that a two-stage approach might not save very much total time in processing major decisions.

A further dimension to the inquiry process is the limited scope for real community involvement in environmental issues. Although supposedly open to the public, inquiries give the impression of being run by experts with the ordinary member of the public ill-informed about inquiry procedure. Although inquiries have the limited status of simply helping an Inspector 'advise' his Minister, who then makes a political decision, inquiries are a judicial part of the democratic process. This idea is enhanced by their adversarial procedure, legal counsel, length and tendency to delay, or excuse delay in, the eventual decision, and by the weight attached to the Insepector's advice – though this may be, and sometimes is, ignored by the Minister (Greenhalgh, 1984; Toms, 1984).

It may be that an unholy alliance is emerging between on the one hand those who resent the unequal position of objectors and the substantial cost to participants in public inquiries and at the opposite end of most planning arguments, those whose initiatives are frustrated or hindered by the present system. Some would argue that it may be cheaper to the national economy to compensate the public for the financial disadvantage of major schemes than to have public inquiries (House of Commons Environment Committee, 1986). Yet the professional and political disquiet at the government's apparent willingness to review public inquiries must be set against widespread dissatisfaction with their use in the context of importance decisions concerning energy or transport policy (for instance, Breach, 1978; Tyme, 1977).

The latest modifications to the inquiry system will do nothing of themselves to improve the *policy* advice from central to local government. Morever, if they do not work, other approaches are likely to be put in the place of inquiries. Parliamentary Bills have been used in the Channel Tunnel case because it appears the planning inquiry system would not guarantee a development timetable to satisfy the French. In other situations where the nature of the proposed development has already undergone public scrutiny and planning policies have identified sites as suitable for development, an alternative to the public inquiry might be a public discussion at which all views about a site are expressed (McRae and Greaves, 1981). However, if public scrutiny of major controversial environmental issues is guillotined too much the cause of the environmental movement may simply be strengthened.

The Problem of Neutrality in State Planning

The growth of the environmental movement represented a challenge to *laissez-faire* economic liberalism (Lowe and Goyder, 1983). Environmental impact assessment requires that the environmental implications of projects are taken into account by planners and politicians before a decision can be reached on whether development should proceed or not, and that market criteria are not the only ones which should prevail.

Some observers argue that the existence of a public planning system, operated through departments of central and local government is sufficient safeguard for environmental values in society; after all, the planning system was brought into existence to protect the environment. Others claim the present system of development plan and development control, initiated under the 1947 Town and Country Planning Act, is too stretched to carry out the research necessary for a proper evaluation of the environmental impact of policies or proposals. Others point to the quasi-judicial role which planners are forced to adopt which makes it difficult for them to consider the wider range of social and political impacts raised by many development plans.

Underlying such viewpoints is a widely held assumption that planners, particularly in local government, are 'competent' (to use the expression in the EC Directive) to carry out EIA. Competence may be judged on technical and ideological grounds. The technical issues have been outlined by Alan Bell (see Chapter Two).

The ideological assumptions are rarely challenged, possibly because one of the fundamental roles for planning officers in government has always been to mediate impartially between the interests of capital and consumption. But planning is essentially a permissive system and has become more so under a market-orientated government ideology. Several implications flow from this assessment. First, the extent to which planners can remain free from incorporation into the State's wider political objectives has become more limited (Knox and Cullen, 1981; Cawson, 1986); second, the potential for neutrality or objectivity on the part of planning officials may have been exaggerated (Reade, 1985); third, planners' own perception and understanding of the importance of environmental problems has been altered by the political environment in which they have to work. There is a further point: it seems that the managerial role of State planning agencies in regulating and initiating development, in public and private sectors, has to some degree been overlooked, both from the point of view of public concerns about government-decided development and the future problems which may arise from today's decisions. It was conventional wisdom to cast doubt on the motivations of big business interests in relation to physical change, to see the conflicts in land use, loss of cherished landscapes and possible social and political upheavals as 'caused' by large-scale urban intrusion in the countryside. Yet in reality the State itself may be as much a villain. For example, there is justifiable public concern when a local authority applies to itself for permission to develop land and buildings. Most developers' decisions are processed by professional planners who thus cannot escape 'blame' for any socially undersirable consequences which may arise from them.

Whose side are the planners on? Those commercial, industrial or development interests which initiate and carry out development, or those individuals and organizations having an interest in conserving the physical environment? It is a difficult question to answer through inspection of formal planning documents. A general distinction can be made between a corporate production-orientated planning process at central government and the county (strategic) planning level and a competititive pluralistic process in district (local) development control planning (see, for instance, Cawson, 1977). Even this theoretical distinction becomes difficult to sustain in the context of the post-1980 Circulars which have required a much more permissive planning regime at district level; and the actual patterns of permissions and refusals in particular locations, appeal decisions and other factors increasingly support organized and individual development interests at the expense of environmental interests. The opportunity for development interests to dominate the Structure Plan process was enhanced by the replacement of the public inquiry (and the right of any individual to be heard) with the 'Examination In Public' consequent upon the changes made to the methods of producing Structure Plans after the 1968 Town and Country Planning Act.

Civil servants play a key role in the affairs of local government planning. Planners working at county level might argue they are free to determine the key issues in Structure Plans, to devise the appropriate techniques, to formulate policies which are relevant to their unique problems and to seek, as best they can, the implementation of their plans in co-operation with other agencies, including district planning authorities. Civil servants help in shaping the process of plan preparation, in selecting the issues of importance, or advising on procedures, in defining the scope of the 'Examination in Public', even to the extent of selecting those 'favoured' interest groups who should be invited to attend (see Barker, 1983, for an exposition of the centralist trend in planning).

When analysing the role of planners in the political process weight must be given to the ideologies of planners, their political masters, and the bureaucratic rules and procedures under which they operate. Planners' training and ideological stances support many elements in a resource conservation approach to land-use management and change: the long- rather than short-term approach to investment decisions, the development of policies for the environment which should be set alongside but not be subservient to non-environmental policies, the co-ordination of public and private interests in land development, the involvement of the public in the process of determining future environment change. But when faced with the restructuring of the British economy and new political realities, both planners and politicians may be reluctant to press for these traditional values for fear of losing jobs or investment or both. As Andrew Blowers says:

> it is obviously difficult for planners to undertake roles which might bring them into sharp and politically uncomfortable conflict with developers, farmers, industry and workers . . . development control, with its rhetoric of conservation, often in reality becomes a surrender to powerful material interests engaged in the protection of amenity, property, profit and jobs. (Blowers, 1984)

Similarly Sandbach argues that the assumption behind EIA is that there is a price to pay for environmental and social impacts:

> the actual price is open to politics and bargaining but as with commodity goods, capital owners and the ruling class are best able to ensure the evaluation process works in their interest. (Sandbach, 1980)

One does not need to be a radical to appreciate the observation that planning is not neutral but rather is influenced by the more powerful groups; it is the implications of what should be done about it which are debatable. Blowers and Sandbach assume a particular ideological position for planning in society: they neatly expose the paradox between a social vision of planning, with its emphasis on questions of equity and social justice, and the more familiar role of planners in practice as mediators between development and environmental interests. The poor and less articulate are unlikely to be well served by the mediating planning system. The effect of many planning decisions is to support the already favoured groups in society. There is therefore some obligation on planners to understand and make known the social consequences of their policies and plans.

Similarly, the planning system has limited value for dealing with environmental issues because it focuses too much on short-term social and economic gain at the expense of long-term environmental stability, and often fails to appreciate the links between the physical and human environment (Park, 1984). Much in-house government planning research seems addressed to solving the short-term information requirements of planners rather than being addressed to the fundamentally long-range effects of the planning for society or environment – themes which are left to academics, environmental pressure groups or concerned politicians to research.

Reade (1985) argues that planners have simply failed to carry out their proper function of policy research:

> they occupy themselves with disguising land use policies, hiding them behind a smokescreen of pseudo-objective techniques and professional mystification . . . the planning professional has successfully defined land use control as an objective necessity, as self-evidently beneficial and therefore *not requiring* any investigation.

Reade thinks that planners cannot be relied on to oppose undesirable social and environmental developments since their professional commitment to the use of land and buildings forbids them to become involved in political debate; if planning control were to be extended into such areas as intensive agriculture or energy policy the techniques and procedures of EIA would obfuscate any proper democratic political debate.

One reason why the consequence of planning policies is rarely addressed at local government level is insufficient manpower and financial resources. Few local authorities budget adequately for policy research of this kind to be carried out, either in-house or through consultancies. Another reason may be that many planners see their role as technical 'experts' and wish to keep away from potentially uncertain research findings which may reveal inconsistencies or weakness in the authorities' policy stance. EIA could easily turn out to be nothing more than a purely technical exercise with planning officers being asked to analyse rather than assess different environmental impacts, leaving judgements about the importance of different environmental impacts to politicians (Ollinger, 1975). Some planners may resist the requirements of EIA to the last believing that existing development control procedures are an adequate means of handling the indirect and direct consequences of a proposal and, most importantly, that the task of putting value judgements on impacts is too difficult for the planning officer (see Baldwin, 1979, for a development of these views).

In the face of these ideological and practical obstacles planners in government are likely to meet the requirements of EIA by independent assessments carried out by planning consultants. EIA effectively complicates the decisions which the planning authority must make by providing a greater range of information, some of it speculative and unquantifiable: planning officers will be reluctant to present this material to elected members if doing so leads to questions about their autonomy or 'neutrality' (for instance, Brooks, 1976; Burton *et al.*, 1983). The use of consultants may not overcome the problem of objectivity – what will be the cost? who will pay? will developers agree to the findings of independent consultants? which consultancy practices should be used? who should decide? are the best known practices the most efficient? who can do the job in the time available? and so on.

Although planners recognize the importance of environmental impact assessment and, one suspects, would wish to see social and environmental values given more weight in public decision making, what they would like to do and what they do in practice are not the same thing (Mayo, 1984). Political realities constrain their decisions. It is local councillors and Ministers who actually make decisions. Nor can it be straightforward to develop a consistent approach to environmental policy at the local scale while being required to facilitate urban development which is in favour by local business interests. Furthermore, much 'planning' is outside the hands of professional planners being carried out by civil engineers, executives' departments etc. Most of these groups have a strong bias in favour of concrete action, and strong prejudices against the 'lower orders' and 'vociferous minorities' which they see as in the way of progress. Finally, we cannot assume that practising planners do not share the values of other members of society: a pro-growth, anti-environmentalist stance may mask a liberal, conservationist and socially aware ethic or vice versa. A split personality may simply be a reflection of the deeper divisions in society between those who favour growth and those who prefer preservation.

The risk for planners of becoming too close allies of the development industry becomes obvious in the context of EIA. While the responsibility for preparing the EIA is strictly with the developer, the planning authority may find it difficult to resist an inadequately prepared assessment, the basis of which they have prior agreed with the developer. If they do not like the EIA, will they refuse the planning application on grounds of inadequacy? and if so, would the Department of Environment uphold a 'planning decision' based on these grounds? Similarly, the planning authority will be the arbiter of what are the *significant* effects of a proposed development: if they have agreed these with a developer their advice will no longer be regarded as objective in the eyes of environmental interests (McDonic, 1985). It seems most likely that planners will incorporate development interests especially in relation to such matters as the form in which the assessment should be prepared, or the issues to be covered, furnishing information and data from its own records if necessary to assist developers. A co-operative approach would at least have the advantage of bringing planners more closely in touch with economic investment decisions, a trend which might be welcomed by some members of the planning profession.

Unfortunately economic decline has pushed governments in Britain to view economic growth as a separate activity to environmental protection, and to believe that changing institutional structures will help bring about increased output while safeguarding the environment. Economic and environmental interests are frequently self-interested, failing to perceive the possibility for greater environmental accountability in economic decisions or the advantages of economic investment for environmental protection. There are major ideological, institutional and financial barriers to bringing different interests together (Johnson, 1983). In this context, the role of State planners is likely to remain ambivalent and ill-suited to the philosophical purposes of the environmental movement.

Environmental Interests in EIA

The strength of the environmental movement has been acknowledged in most countries . . . by the creation of institutional structures designed to ensure the incorporation of such values within the decision-making processes of both public and private agencies. (Chapman, 1982)

This statement is probably a fair summary of the position seen by industrialists facing the need to overcome the environmental objections raised by organized and individual interests. The existence of land-use planning controls, pollution control measures and other aspects of government policy must seem at times a rather formidable array of instruments for curbing the activities of developers. But is this how the environmental lobby perceives the role of institutional structures? If the planning system is essentially permissive to what extent can environmental values be incorporated into decision making? Indeed, have environmental politics become an irrelevancy under a planning system which for most of the time incorporates development interests? Such questions raise issues of international importance (Knoepfal and Watts, 1983).

Answers must remain speculative but we gain some insight into them by examining the opportunities for influencing decisions which are currently open to the environmental groups, and by asking whether EIA is likely to help to enlarge these. We can also find out whether, how and when the public will be consulted when the new EIA procedures are introduced.

At national level the participation of environmental groups in decision making is associated with parliamentary debate, especially when backbench MPs table questions or sponsor Early Day motions. Recent examples of environmental issues covered in parliamentary debate include: the dangers of dispersed urban development and its possible impact on farmland and landscapes in the outer South East, the impact of the M25 motorway, the siting of a third London airport at Stansted and the question of a fifth terminal at Heathrow, and the route of the Okehampton By-Pass. The new Select Committees do challenge government policy. Even without obviously shifting policy they may have value in bringing environmental concerns to the attention of government and the public as well as educating the environmental specialists in how to present their case. The political parties set up their own environmental groups and increasingly recognize the electoral importance of the 'green' vote. But at national level the link with political parties is weak and corporate interests in business, industry and the trade unions tend to be more entrenched in government policy than are environmental organizations (Lowe and Goyder,

1983). However, there is little evidence of lessened environmental pressure group activity at national level despite any difficulties such groups may face in having their voice heard. The Council for the Protection of Rural England and the Town and Country Planning Association have special interests in promoting environmental protection and public planning and have good channels of communication with the media. It is difficult to judge when lobbying is successful where central government planning decisions are involved because of the codes of secrecy built into Departments of State. A combination of political representation, pressure group lobbying, civil disobedience and public censure have helped bring about some short-lived project reappraisal (for instance, in respect of proposals for the underground disposal of nuclear waste) but for the most part government has been more inclined to give priority to projects of whatever kind which it believes will stimulate the economy, and to overrule environmental objectors if it suits it. Thus, the national motorway programmes have been delayed but not hindered, the recent decision to expand Stansted airport went against considerable public opposition, and in the last months of 1985 the government made it clear that it did not wish to see the Channel Tunnel project delayed by a local planning inquiry.

Will new EIA procedures improve the lot of objectors, either by giving more weight to the economics of resource decisions, or by directly allowing environmentalists a greater say in decisions? One immediate problem is that the application of environmental impact assessment to national policies has never been seriously considered. There is no guarantee that environmental groups will be any stronger especially while national policies are strongly development-orientated. Morever, statutory EIA will be directed to improving the quality of local rather than central government planning. Some national policy shifts may give added encouragement to local pressure groups; for instance, the potential for lobbying against development in specially designated areas like National Parks was enhanced in 1981 when land-use planning controls were relaxed in non-designated areas, thus effectively producing a two-tier system of environmental protection in England and Wales, with the non-protected areas covering some 70 per cent of the total land area (Council for Protection of Rural England, 1981). The advent of EIA could strengthen the hand of quasi-autonomous agencies like the Countryside Commission or Nature Conservancy Council when these bodies are consulted about specific environmental threats, insofar as they would be able to urge local planning authorities to use EIA in Annex 2 cases; in this sense other environmental groups could feel their interests were being safeguarded rather better than at present. The principle of EIA could also be extended to such questions as the social and environmental impact of designation policies, thus providing a better informed basis for any public inquiries which occur when new designations are proposed. It is interesting to note that the decision to designate the northern Pennines an Area of Outstanding Natural Beauty was taken without any formalized use of EIA and that after many years of delay in arriving at a decision the government felt it necessary to hold a public local inquiry into the proposal – how much speedier the decision process might have been if the Countryside Commission had prepared a thorough EIA in the first place.

More worrying is the standard administrative response to calls for greater environmental accountability in decision making: the circumventing of delays by eliminating consultation. The British government's reluctance to take on board the EC Directive may be explained by its fear of delay in arriving at important national decisions plus a fear that EIA might halt or hinder measures it favours; yet delays are likely to be no greater with EIA than without and there is the chance that decisions will be more cost effective in the long term (Fisher, 1981). It is not consultation which causes delay as much as the absence of policy against which to test particular projects and the lack of policy as opposed to project evaluation. Democratic participation in strategic planning decisions is necessary and could be improved if EIA were applied to major projects in the future.

At present the public is involved in planning decisions at various stages in the decision process. Comparative data from the United Kingdom and United States of America suggest that participation is most effective in small-scale developments and progressively less effective as the size of project increases: citizen participation has little effect on public infrastructure of housing schemes and almost no effect, despite much formal participation, on national projects like motorway (Johnson, 1984). Based upon this research it seems likely that decisions on big projects, which will be the only ones subject to formal EIA procedures, are unlikely to be swayed by any environmental views expressed by the public.

Unfortunately the EC Directive is notably vague about participation procedures: will environmental groups as well as developers be consulted at an early stage in drawing up the check list of issues to be included in the EIA? By involving the public prior to a decision on a planning application the time taken on any subsequent appeal could be reduced since the main arguments, and the information on which they are based, will be available to all sides in a dispute. Will the same groups and interested members of the public be allowed to see the results of the EIA prior to a final political decision and if not, how will their views be canvassed by the local planning authority?

Despite its faults the inquiry system is still a useful mechanism for examining particular projects, provided policy evaluation has been carried out. Those who argue against the delay and cost of inquiries ignore the improvements which the greater use of EIA could bring to them: some measure of pre-inquiry agreement on the key issues, information on a comprehensive range of impacts, a developer's evaluation which is open to public scrutiny. If the involvement of environmental groups in the preparation of EIAs could be guaranteed then it might be politically acceptable for a government to reduce the length of time for inquiries, with a view to saving public money and financing impecunious objectors. This kind of reform accords with Lord Young's 'Lifting the Burden' initiative, but the use of a statutory fixed time limit for a final ministerial decision (from the initial appeal on a planning application) would probably be impractical in the case of Annex 1 developments which the government agrees require complex and systematic assessment and possibly difficult to impose in the case of the more numerous Annex 2 developments. A final point is that were the public inquiry system ever to be abandoned, environmental groups might find themselves faced with an even less democratic mechanism for dealing with major projects: namely the use of hybrid Bills in Parliament which may not be subject to international or national or local environmental impact assessments within the terms of the EC Directive.

There remain fundamental differences between those planners and developers who view EIA as a management tool, a means of making more informed decisions (whatever the outcome) and those who argue that the potential of EIA lies in the opportunity it creates for the incorporation of environmental values among decision makers. There need be little conflict between doing EIA at a time of more relaxed planning if the purpose of EIA is simply to bring about rather better designed schemes without actually restraining development. On the other hand if government agencies are to incorporate environmental values into their decisions there would seem little point in carrying out EIA if it is simply an additional bureaucratic procedure removed from the substantive concerns of the environmentalists. In those countries which have imposed formal EIA procedures the efforts of the environmental movement have been dissipated in challenging complex documents rather than in questioning agencies' powers or responsibilities to make environmentally sensitive decisions (Fairfax, 1978; Burton et al., 1983). The danger is that EIA will not lead to the expression of environmental factors in planning decisions, but rather be used to 'resolve' conflict and legitimate the State's real interest in securing economic development. Brian Wynne (1980) might have been speaking of EIA when he warned:

> no process, however theoretically ideal and well designed can compensate for more general problems such as . . . a widespread inability to articulate social choice and concerns, or a lack of political sensitivity to changing currents of feeling in society.

EIA of itself is unlikely to protect the environment although environmental interest groups may attempt to use EIA for this purpose. The ultimate decision whether or not to proceed with a development will depend on economic and political as well as environmental factors. Environmental groups will remain weak where local government is strongly development-orientated. In Britain the present government's industrial strategy places priority upon employment generating development. Such policies are likely to colour the political decisions on applications for planning permission whether an EIA has been carried out or not, and often irrespective of the role played by environmental groups in the decision process. We cannot know how much influence environmental groups will enjoy at a time of increasing government concern with material welfare but if the view that 'environment' is a luxury society cannot afford becomes commonplace the confidence of environmental interest groups in EIA as both an assessment procedure and a philosophy could quickly erode.

Toward an Environmentalist Approach

A distinction should be made between the limited effectiveness of environmental interests in shaping public decisions and the emergence of environmentalism as a social force and political philosophy. Despite conditions of high unemployment and continued economic uncertainy in Britain there is no slackening of public enthusiasm for environmental protection. A recent opinion poll showed that pollution and resources depletion were perceived as important issues affecting people's daily lives, although unemployment, inflation, law and order were immediate national concerns (Johnson, 1983). In the longer term there are several pointers to continued environmental interest in society which may prove resistant to economic cycles: increased questioning of the likely benefits of particular kinds of economic development (e.g. high-technology industry or defence industry) especially in relation to their impact on future levels of employment and unemployment and their potentiality for redistributing wealth from richer to poorer groups; greater popular acceptance that a pleasant and safe environment are important factors in general standards of welfare, alongside housing and health; and a diffusion of both these concerns among all social classes (see Lowe and Goyder, 1983 for a full discussion).

Moreover, the factors which prompted the environmental movement in the 1960s and 1970s remain alive, albeit in a different guise, in the 1980s. Such issues as the spread of urban development in the countryside, or the more intensive use of farmland still reflect

> a widespread debate about the quality of life in post-industrial conditions . . . and the contribution that development, and the economic growth it is supposed to generate, might make to improving these conditions. (Wandesforde-Smith, 1980)

Furthermore, some 'green' groups seek a fundamental change in political attitudes to the growth cycle of Western industrial economies, urging a return to the ecological principles which found favour in the mid-1970s (*Ecologist*, 1973; Club of Rome, 1974). Others are prepared to use militant tactics to pursue their environmental goals in Europe and elsewhere (Pilat, 1980). The rise of Green and Ecology politics seems to stem in part from a feeling that individuals and governments have been morally irresponsible towards the interests of the next generation; in part from an expression of anti-science and technological values and a concern about the pace of scientific change; and perhaps in part from a belief that political and economic systems are ill-adapted to handle problems of economic change (Harvey and Hallett, 1977). However implicit these feelings, there is little doubting the importance of environmental issues nor their public expression. Future government responses are likely to be influenced by any long-term changes in social values especially if they find political expression. In Britain it is already obvious that some popular green ideals and objectives are being internalized within most of the political parties.

Yet the effectiveness of more formal EIA procedures is likely to be reduced in practice by the changed planning system, and the uncertainties surrounding its adoption in United Kingdom planning departments. The reason for this does not lie with the lack of technical expertise within local planning agencies, but rather with prevailing government policy stances which are strongly pro-growth, an issue which has been raised elsewhere in the context of Australian experience (Rees, 1982). The growing environmental movement presents a considerable challenge for future governments: how can social and environmental values be incorporated into public and private sector decisions whilst maintaining legitimate interests in economic change? There is a fundamental dilemma between society's need for some form of social control over industrial development and the interest of the development lobby in avoiding excessively burdensome regulation. If the planning system is failing to care for the public interest in environmental protection the question is then raised: under what conditions could the politics of planning be made more responsive to environmental concerns?

Government Priorities.   Economic issues often seem to push governments to a simple minded concentration on consumption and economic growth or to a mistaken belief that changing institutional structures will bring safeguards for the environment. The real ideological, institutional and financial barriers to the integration of environmental and development interests have been mentioned before (Johnson, 1983). The conversion of governments, as distinct from politicians, to environmental concerns is likely to be a slow and halting one while the environmental ethic is seen to oppose conventional economic wisdom. While environmental improvement is seen as an inhibitor of wealth creation, governments may choose to argue that 'only the affluent can afford to be environmentalists' (Holmes, 1976). But could not an alternative case be made for regarding environmental protection as a form of wealth creation? It is a false economy to ignore the environment. Environmental protection measures have a significant pay back – the costs of pollution control are less than those of pollution – a point which had been dramatically demonstrated by the British government's late and expensive response to the problem of acid rain. Hence a small proportion of Gross National Product to meet the costs of all environmental management would be more than met by the damage or risk forestalled as a result of this action. The direct costs of environmental action may seem high but the costs of having to respond later to unforeseen problems is rarely quantified. Savings to local and national economies arising from the avoidance of long-run damage outweigh the short run costs of using EIA procedures (see Dean and Graham, 1979; House of Lords, 1981).

Environment as a Strategic Policy Issue. The opportunity is presented when major projects come under public scrutiny. In Britain there may be rather few of these in future, excepting the Severn Barrage and the South Warwickshire coalfied prospect. But the introduction of EIA could be extended to create mechanisms for environmental participation and accountability in the policy, plan and programme making of government departments (an excellent critique of the possibilities is given by Wandesforde-Smith, 1980). Some kind of strategic planning framework would seem necessary for dealing with the impending Annex 2 list of projects which will impinge upon (at least indirectly) policy arenas such as extractive minerals, various kinds of manufacturing industry and infrastructure related projects including urban development.

A regional tier of government planning, possibly groupings of existing planning authorities, would be in a good position to carry out, in the context of policy formulation and review, an assessment of the strategic implications of government policies for say toxic waste, oil refineries, new settlements or hyper-markets. District authorities could then be encouraged to include in their Local Plans any mention of those policies and developments which they would like to see covered by an environmental impact assessment, even though the onus of producing an EIA would still fall upon the developer. The question of whether such policies or projects should require an EIA would be adjudicated by the DOE, as would the form of modifications to the plans of those authorities failing to identify relevant policies. Without some system of policy review, of the kind carried out in the monitoring or alteration stage in structure planning, it will be difficult to know how central government will assess the importance of any subsequent EIA submitted by a developer should a case be taken to planning appeal.

Unfortunately the opportunity for policy review of this kind depends on changing official attitudes toward policy formulation and review. The breadth of policy aspects would need to be wider than anticipated by the EC Directive on EIA. Many pressures on the local environment do not require permission to proceed, for example changes within the industrial Use Classes. Morever government departments and statutory undertakers are excluded from planning control (Booth, 1984). The mechanisms for policy monitoring already exist and would need to be extended to cover those policies on energy and transport for instance which are the province of separate government departments. Formulation of national policy priorities for the environment would require more overt recognition of the informal lobbying relationships that exist between say development interests and environmental pressure groups in the national arena and would imply a clearer hierarchy of decision making between the centre, regions and localities whcih does find favour with some political parties. Finally, the greater involvement of Parliament in policy issues might best be achieved by building on the research expertise supporting the main parties. The parliamentary Select Committee enjoys an improved level of expertise and greater political respect than seemed likely at one time – it could be used as an environmental policy-reviewing institution in selected cases (O'Riordan, 1982). The all-party support for the Select Committee on the nuclear power industry and the Committee reports on Green Belts and Housing Land showed how environmental values can be given greater expression at national level.

New Land Planning Policies. These present the opportunity for greater incorporation of environmental interests in government decision making. The social problems presented by a more dispersed and mobile society are improperly understood, nor is the nature of the public response clear. Studies of post-war urbanization in Britain suggest that urban containment, with its twin goals of countryside protection and planned concentration of new development (as in New and Expanded Towns) is becoming out-of-date. But what to put in its place? The need for clearer thinking about the future of land and settlement is indisputable given the projected changes in the agricultural industry. Farmland which may be surplus to requirements for food production may open the way for new kinds of environmentally sensitive developments such as forestry, organic farming, tourism and small-scale housing (Dower, 1986). Already there are new ideas for dealing with the issue of agricultural overproduction (MAFF, 1986; Royal Society of Nature Conservation, 1986). Fewer restrictions on new housing building would be favoured by the housebuilders but the impetus this would give to continued population dispersal would be an insufficient public goal on its own without the recycling of land, equity of provision for different social groups, and partnership between private and public interest in new housing (Royal Town Planning Institute, 1986). One way of legitimizing social values in government policy would be via a return to *advocacy approaches*. The United States and Canadian governments in part tackle the issue by establishing independent panels composed of government agency and community representatives charged with the job of defining and guiding social impact assessments (SIAs), soliciting public comment and making recommendations (Rohe, 1982). The purpose of this approach is to involve the community in the actual preparation and assessment of impacts. The carrying out of an SIA provides a good means of extending participation and educating the public. For it to be effective legislation should specify when and how SIAs should be carried out with provision for stopping projects which result in severe social impacts. Funding, technical assistance and access to information are necessary. Because SIA methods of this kind involve an apparent shift in power away from the planners and developers to the community they are likely to be resisted by government agencies; obviously they require political support if they are to get off the ground. Yet advocacy planning has some potential advantages: it could be used in the context of large projects and usefully be applied to a greater number of specific developments which are small in scale, occur frequently and impact on local communities; examples include the activities of mineral developers (such as sand and gravel extraction) or regional housebuilding companies. The use of SIA approaches has been comparatively neglected in Britain. Yet they could provide a means to complement the more formalized methods of EIA which seem most likely to bolster the interests of the large at the expense of the small developers and to exclude the legitimate concerns of local communities.

Summary

EIA may be viewed by technical rationalists as simply an administrative procedure for ensuring greater weight is given to environmental factors in decision making. Important though this is, many commentators would admit that the purpose of using EIA extends to bringing about changes in attitude toward the need for and design of new development (Bidwell, 1985). The big question is *will* environmental considerations be adequately integrated into land-use planning given the prevailing ideologies of politicians and planners. This chapter's rather gloomy prognoscis is that, despite a growing political consciousness about the environment, politicians and bureaucrats will continue to allow economic priorities to outweigh those of the environment in public decision-taking. The market orientated values of government will sit uneasily alongside the emerging environmental values in society. National government pressure to regenerate the economy may continue to override local interests and increasingly seem out of step with the expression of social and environmental concerns.

In this uncertain political context EIA is likely to have nothing more than a peripheral or at most marginal impact on the quality of most planning decisions. Three issues, in particular, are likely to limit the value of EIA in the short term: first, more permissive national government attitudes to the environment; second, political realites which constrain the 'neutrality' of local planning decisions to arbitrate in favour of the environment; and third, lack of public confidence that legitimate environmental concerns will be taken into account in public decision-making. For EIA to exert a more potent role in the long run would seem to require nothing less than a complete rethinking of national planning goals. Clearer government priorities for the environment and effective strategic planning are necessary. New land policies and community-based approaches should be tried if the environmental values which underpinned the introduction of EIA are to be properly realized.

REFERENCES

Armstrong, J. (1985) *Sizewell Report: A New Approach for Major Public Inquiries.* London: Town and Country Planning Association.

Baldwin, K. (1979) Environmental impact assessment and development control, in Herington, J. (ed.) *The Role of Environmental Impact Assessment in the Planning*

Process. Loughborough: Institute of British Geographers and Department of Geography, Loughborough University, pp. 70–78.

Barker, A.P. (1983) Public participation and the central control of structure planning, in Bristow, M.R. and Cross, D.T. (eds.) *English Structure Planning*. London: Pion.

Bidwell, R. (1985) The gap between promise and performance in environmental impact assessment: is it too great? Paper presented to conference on The EIA Directive: Implications for Practice and Training, 26 September. Manchester: Department of Town and Country Planning. Mimeo.

Blowers, A. (1984) Planning's crisis of credibility. *Town and Country Planning*, 53 (12), pp. 345–8.

Booth, A.G. (1984) The planning component, in Roberts, R.D. and Roberts, T.M. (eds.) *Planning and Ecology*. London: Chapman and Hall.

Breach, I. (1978) *Windscale Fallout*. Harmondsworth: Penguin.

Brooks, E. (1976) On putting the environment in its place: a critique of environmental impact assessment, in O'Riordan, T. and Hey, R.D. (eds.) *Environmental Impact Assessment*. Farnborough: Saxon House.

Burton, I., Wilson, J. and Munn, R.E. (1983) Environmental impact assessment: national approaches and international needs. *Environmental Monitoring and Assessment*, 3 (2), pp. 143–7.

Cawson, A. (1977) Environmental planning and the politics of corporatism. University of Sussex, Department of Urban and Regional Studies, Working Paper 7.

Cawson, A. (1986) *Corporatism and Political Theory*. Oxford: Blackwell.

Chapman, K. (1982) Environmental policy, industrial location in the United States, in Flowerdew, R. (ed.) *Institutions and Geographical Patterns*. London: Croom Helm, pp. 141–68.

Clark, M. (1978) Environmental impact assessment: an ideology for Europe. *Town and Country Planning*, 46 (9), pp. 395–9.

Club of Rome Report (1974) *The Limits of Growth*. London: Pan Books.

Council for the Protection of Rural England (1981) *Planning – Friend or Foe?* London: CPRE.

Council for the Protection of Rural England (1983) M40 Public Inquiry: proof of evidence of A.R. Long. London: CPRE.

Dean, F. and Graham, G. (1979) The Application of Environmental Impact Analysis in the British Gas Industry. Paper presented at the Economic Commission for Europe Seminar on Environmental Impact Assessment, Villach, 24–29 September.

Department of Environment (1982) *Development Control – Policy and Practice*, Circular 22/80. London: HMSO.

Department of Environment (1981) *Development Plans*, Circular 23/82. London: HMSO.

Dower, M. (1986) Ferment of ideas for a rural future, *Town and Country Planning*, 55 (12), pp. 339–41.

Ecologist (ed.) (1973) *A Blueprint for Survival*. London: Penguin.

Fairfax, S.K. (1978) A disaster in the environmental movement. *Science*, 99 (4330), pp. 743–8.

Fisher, A.C. (1981) *Resource and Environmental Economics*. Cambridge: Cambridge University Press.

Garner, J.F. and O'Riordan, T. (1982) Environmental impact assessment in the context of economic recession. *The Geographical Journal*, **1481** (3), pp. 343-61.

Goodchild, R. and Munton, R. (1985) *Development and the Landowner: An Analysis of British Experience*. London: George Allen and Unwin.

Goldsmith, E. (1972) *A Blueprint for Survival*. London: Stacey.

Greater London Council (1985) *Erosion of the Planning System*. London: GLC.

Greenhalgh, G. (1984) The Sizewell Inquiry – is there a better way? *Energy Policy*, **12** (3), pp. 283-7.

Gregory, R. (1971) *The Price of Amenity*. London: Macmillan.

Haigh, N. (1983) The EEC Directive on environmental assessment of development projects. *Journal of Planning and Environmental Law*, September, pp. 585-95.

HMSO (1986) Command Paper 43. *Planning appeals, Call-Ins and Major Public Inquiries* – the government's response to the fifth report from the environment committee, Session 1985–86, Cm 43. London: HMSO.

Harvey, B. and Hallett, J.D. (1977) *Environment and Society – An introductory Analysis*. London: Macmillan.

Herington, J. (1984) *The Outer City*. London: Harper and Row.

Holmes, N. (ed.) (1976) *Environment and the Industrial Society*. London: Hodder and Stoughton.

House of Commons Environment Committee (1986) *Planning Appeals, Call-Ins and Major Public Inquiries*. London: HMSO.

House of Lords Select Committee on the European Communities (1981) Environmental Assessment of Projects, House of Lords Paper 69. London: HMSO.

Johnson, B. (1983) *The Conservation and Development Programme for the UK*. London: Kogan Page.

Johnson, W.C. (1984) Citizen participation in the UK and USA. *Progress in Planning*, **21** (3) pp. 153-221.

Knoepfal, P. and Watts, N. (eds.) (1983) *Environmental Politics and Policies: An International Perspective*. Frankfurt: Campus Verlag.

Knox, P.and Cullen, J. (1981) Planners as urban managers: an exploration of the attitudes and self-image of senior British planners. *Environment and Planning* A, **13** pp. 855-98.

Lawson, P. (1982) A 'regrettable necessity' – the decision to construct. *Hydrobiologia*, **88**, pp. 19-26.

Liberal Party Environmental Co-ordinating Group (1986) *Survival: The Liberal way to an environment for the future*. Hebden Bridge: Hebden Royal.

Lowe, P. and Goyder, J. (1983) *Environmental Groups in Politics*. London: George Allen and Unwin.

Ministry of Agriculture, Fisheries and Food (1986) *Diverting land from cereals*. London, September.

Mayo, J.M. (1984) Conflicts in roles and values for urban planning. *Journal of Architecture and Planning Research*, **1** (1) pp. 67-77.

McDonic, G.F. (1985) Environmental impact assessment, implications for practice: planning authority's view. Paper presented to conference on The EIA Directive: Implications for Practice and Training, 26 September. Manchester: Department of Town and Country Planning. Mimeo.

McKay, D.H. (ed.) (1982) *Planning and Politics in Western Europe*. London: Croom Helm.

McRae, M. and Greaves, J. (1981) Nigg Bay petrochemical developments: a successful alternative to a public inquiry? *Scottish Planning: Law and Practice*, **2**, pp. 8–10, 19.

Ollinger, W. (1975) Environmental impact – a political process, in Hutchings, B., Forrester, A., Jain, R.K., and Balbach, H. (eds.) *Environmental Impact Analysis: Current Methodologies and Future Directions*. Illinois: Department of Architecture, University of Illinois.

O'Riordan, T. (1976) Beyond environmental impact assessment, in O'Riordan, T. and Hey, R.E. (eds.) *Environmental Impact Assessment*. Farnborough: Saxon House.

O'Riordan, T. (1982) Institutions affecting environmental policy, in Flowerdew, R. (ed.) *Institutions and Geographical Patterns*. London: Croom Helm, pp. 13–140.

O'Riordan, T. and Sewell, W.R.D. (eds.) (1981) *Project Appraisal and Policy Review*. New York: Wiley.

Park, C.C. (1984) EIA – panacea for environmental management? *Productivity*, **24** (4), pp. 431–7.

Pilat, J.F. (1980) *Ecological Politics: The Rise of the Green Movement*. Beverly Hills: Sage.

Reade, E. (1985) Planning's usurpation of political choice. *Town and Country Planning*, **54** (6), pp. 184–6.

Rees, J. (1982) discussant in Garner, J.F. and O'Riordan, T. Environmental Impact Assessment in the context of economic recession. *Geographical Journal*, **148** (3), pp. 343–61.

Rohe, W.M. (1982) Social impact analysis and the planning process in the United States: a review and critique. *Town Planning Review*, **53** (4), pp. 367–83.

Royal Society of Nature Conservation (1986) *The Countryside Tomorrow – a Strategy for Nature*. Lincoln: RSNC.

Royal Town Planning Institute (1986) *The Challenge of Change: the planning response to social and economic change*. Leamington Spa: Pallados Limited.

Sandbach, F. (1980) *Environment, Ideology and Policy*. Oxford: Blackwell.

Toms, M. (1984) Some reflections on the Stansted Inquiry. *The Planner*, **70** (5) pp. 22–24.

Tyme, J. (1977) *Motorways versus Democracy*. London: Macmillan.

Wandesforde-Smith, G. (1980) Environmental impact assessment and the politics of development in Europe, in O'Riordan, T. and Turner, K. (eds.) *Progress in Resource Management and Environmental Planning*, Volume 2. New York: Wiley.

Watkins, L.H. (1981) *Environmental Impact of Roads and Traffic*. New Jersey: Applied Science.

Wynne, B. (1980) Windscale: a case history in the political art of muddling through, in O'Riordan, T. and Turner, K. (eds.) *Progress in Resource Management and Environmental Planning*, Volume 2. New York: Wiley.

Chapter Nine

Environmental Impact Assessment In Planning Practice: Will It Work?

GEORGE McDONIC

The requirement for an environmental impact assessment in relation to major developments has never been in greater need than it is today. Rapid urban and industrial change brings both direct and indirect consequences for the environment and society and requires an effective response from policy makers. Planning authorities and others interested in land-use matters have considered it essential that the full effects of major developments should be properly investigated and considered at the time permission to develop is sought, and they were therefore, heartened to know that the Council of European Communities issued a Directive on Environmental Impact Assessment in June 1985.

Some planners view EIA with scepticism whilst others see it as offering an opportunity for a better approach to those types of development which are most likely to have a significant effect on the environment and also to be most controversial to members of the public and elected Councillors.

Many people working in town and country planning consider that the Directive will merely consolidate procedures and practices which have been undertaken in respect of major developments for a long period of time and at the same time feel that, because of the restricted application of the Directive, this may have an adverse affect on their present working practice.

The Department of Environment's view is that the EIA will improve the quality of decision making and assist an industrial developer in his consideration of the environmental effects of a project.

The Directive has been received with apprehension since it was not known whether its impact would be as telling as an earthquake or a tremor. Now that the British government has published its consultation papers on the implementation of the Directive, it seems clear that its effect on the British planning system is more likely to be that of a mere tremor. It is significant that the British government sees the Directive and its work being closely linked into the British planning system and not to be considered separate and detached from it, and for that reason it must be acknowledged that this will be helpful to the planning authorities, the developers and the public alike.

The Directive states that it will be necessary to provide an environmental impact assessment of all categories of development falling within Annex I, and the number of categories is limited to nine, most of which are exceptional types of development. (Commission of the European Communities, 1985). One estimate has suggested that there may be not more than twenty proposals for development which fall within Annex I in any one year throughout Great Britain. The Directive provides that Annex II, which is a much more expansive list of developments, shall be made the subject of assessment if the Member State considers that their characteristics so require. The British government does not foresee that it will be necessary to make the carrying out of formal environmental impact assessments mandatory in such cases, but it is proposed that in relation to projects falling outside Annex I that the appropriate Secretary of State should have the power to direct that an assessment should be carried out in any particular case. The British government's view is that it is not appropriate for the local planning authority itself to seek to require developers to undertake formal environmental impact assessments for projects other than those listed in Annex I.

Reference has already been made to the likelihood of there being a very small number of planning applications falling within Annex I of the Directive; this is particularly so since three of the project types are not normally approved under the Town and Country Planning Acts. One of these projects is that of 'motorways and express roads', a category that might be expected to produce the greatest number of proposals falling within Annex I. It is fundamental that the implementation of the Directive should be undertaken in respect of these types of development in the same way as those types falling within the normal town and country planning procedures. The government has said that it is intended that this should be the case, although no proposals have been prepared as yet. Unless there is a common scheme for submission, right of objection and analysis of all environmental assessments required in respect of all categories of development within Annex I of the Directive there will be distrust of the system and it will fail to achieve the objects it purports to achieve.

The present intentions of the government with regard to implementation seem satisfactory apart from the point mentioned above and it is sensible to lock environmental impact assessment into the present planning system. The planning application will need to be accompanied by an EIA statement and it will not be valid unless this is done. Whilst there will be no prescribed form which the environmental impact assessment should take, the minimum requirements of Annex II of the Directive will apply. It is expected that before an environmental assessment is prepared a prudent developer will seek discussions with the local planning authorities and other relevant authorities to ensure the basis of the assessment will cover those matters likely to be of interest to the parties concerned.

Failure to have these early consultations will leave the developer open to facing a requirement from the local planning authority for further information under the terms of the provisions of the General Development Order, 1977. As has already been indicated, where environmental impact assessments have been prepared in the past there has usually been close co-operation between the developer and the local planning authority, and the EIA has covered the requirements of the authority. For the future it is expected the prudent developer will continue with this practice and it is not anticipated that difficulties about the content of the impact assessment will arise providing this relationship between developers and authorities continues.

It is at this point public sector planning practitioners begin to feel anxious, particularly those who, in the past, have required environmental impact assessments to be prepared to accompany planning applications on a range of proposals falling outside the terms of Annex I of the Directive. There is little doubt that when such environmental assessments have been carried out they have been beneficial to the public and the planning authority alike and many are worried that the inflection of the government's attitude as expressed in its consultation paper on the Directive will lead to a reluctance on the part of many developers to respond to requests by planning authorities for an environmental assessment to accompany a planning application.

There are many examples where responsible developers have provided excellent environmental assessments with their planning applications without any prompting by the local planning authority. This is particularly so in southern England where planning permission has been sought to prospect for oil and gas (Carless Exploration Ltd, 1984, British Petroleum Development Ltd, 1986). Inevitably these submissions have assisted planning authorities in making their decisions and have allowed the public a much better opportunity of understanding the significance and ramifications of the proposals. Those planning authorities facing this type of application will wish to continue to receive environmental assessments notwithstanding the statements of the government.

It is significant the government has recognized that the requirements of the Directive can be met within the context of the present system, and in its words 'without imposing significant new burdens on either developers or planning authorities'. It would be difficult to envisage how it could be possible for environmental assessment to be carried out and analysed outside the sophisticated planning system that exists in Great Britain, and this acceptance that the work arising from the Directive should be grafted onto the planning system is satisfactory. Whether or not it will impose significant new burdens only experience will tell. The involvement of the public is likely to be a significant aspect of this work.

The government has also charged the developer with the responsibility for preparing the environmental impact assessment but reminds local planning authorities of their duties to assist in this work. It seems likely that local planning authorities are likely to have some basic information which will be helpful in the preparation of the assessments, and this information should be made readily available to the developer. It is in the interests of the local planning authority to ensure that the environmental assessment is as complete as it is possible to achieve; therefore it has a strong responsibility to make available such information that is within its knowledge to ensure this is done effectively.

Non-Statutory Bodies

It is recognized that this procedure is not to the liking of non-statutory bodies who feel that the developer and the public authorities will have agreed the basis of the environmental assessment without their participation and input. Because of this the non-statutory bodies feel that they will be at a disadvantage and that the assessment may not cover essential aspects from their point of view, and the only time they will be able to seek to rectify this position will arise when the planning application is submitted accompanied by the environmental impact assessment. They judge that this is likely to be considered too late to require further information, and in any case such bodies have no power to require additional information. In these circumstances they are left with only two remedies, neither of which they consider to be satisfactory. Firstly, to ask the local planning authority to require further information from the developer under the terms of the General Development Order 1977 or, secondly, ask the Secretary of State to call in the application for his decision in the hope that the 'missing' information can be required by the Secretary of State at the Public Inquiry or be elicited through cross examination at the Inquiry. Both these remedies have strong disadvantages to the non-statutory bodies, for in the case of requiring the local planning authority to ask for additional information this must be done within a very short time scale and means the non-statutory body must analyse the planning application and the environmental impact assessment and then determine any inadequacies. Having done that it must persuade the local planning authority to take action very quickly, and this may prove to be difficult because of the time allowed by the General Development Order to require further information by the local planning authority. This time constraint would work against the non-statutory body or individual who was dissatisfied with the content of the environmental assessment. With regard to the other option this may be more attractive if the Secretary of State calls in the application, but the non-statutory objector faces the prospect of arguing before the Secretary of State not only that he considers the environmental impact assessment is inadequate but also that the local planning authority should have found it deficient.

The EC Environmental Assessment Directive Working Party (1985) favoured a different pre-application procedure in respect of Annex I proposals which may well have overcome these difficulties by requiring a formal notification of the proposal and intention to prepare an environmental impact assessment and this to be made known publicly some set time before the actual submission of the planning application. It had been thought that this procedure would not only let the general public know of the intention of a planning application in respect of a particular development but would allow for non-statutory bodies and individuals to indicate their expectations of the content of the environmental assessment. Ministers concluded that the formal notification requirements proposed by the Working Party could not be justified.

Past experience has shown that where environmental impact assessments have been prepared the interests of non-statutory bodies have been taken into consideration, but whether this will happen in the future, now the government has prescribed in this matter, remains to be seen. Fortunately most of the developers likely to be covered by Annex I are sufficiently responsible to be responsive to all essential inputs to an assessment and hopefully will do so. It is also argued that it will be very difficult to keep secret that a major project is being planned and therefore those who wish to ensure that the environmental impact assessment embraces the issues of concern to them will be able to make representations at an early enough stage to the developer and the local planning authority.

The decision of Ministers on this point is understood, but at the same time it seems that experience will be the only judge of the effectiveness of their action. Because of the small number of cases involved it may prove not to be too difficult a point. It may however be a very important issue in relation to those applications falling outside the planning system.

Public Involvement

The whole question of public participation in environmental impact assessment work seems to have received scant treatment, particularly for the man in the street. What opportunities does he get, what rights does he have, and what remedies are available to him if he does not find the assessment meeting his objectives?

It has just been explained how the environmental impact assessment will be submitted with the planning application in respect of those types of development falling within the present town and country planning legislation. This is the first time the public will know of the application since it is likely that it will require an advertisement, but in any case the local media is certain to pick it up and give wide coverage of the proposals. At this stage the planning application and the environmental impact assessment will be available for public inspection at the local authority offices. It is likely that the assessment will be a bulky document and will require considerable time for reading to assess its implication. A copy should be available in the local library; nevertheless it will require considerable research to establish all the contents. It would be helpful to the lay person if an abridged version were available setting out the essential components, and this could supplement the non-technical summary of information that is required under the terms of the Directive. Any individual will have the right to inspect the planning application and the assessment which accompanies it and to give views to the local planning authority. Beyond this his rights are very limited, particularly if all the proper procedures have been followed. If the Secretary of State calls in the application for his decision and holds a public inquiry he will have the opportunity to object to the proposal and assessment and to appear at the local inquiry.

Some question whether this is adequate protection for an individual. The proposals of the Department of Environment Working Party to require a pre-submission notification of the application to the local planning authority would have given the individual a possible opportunity to influence the content and quality of the EIA. The argument of the government was that as the environmental impact assessement is to be grafted onto the existing planning system (where it applies) therefore the present position of lodging objections will continue and this should prove adequate. However, in the case of major proposals where an environmental assessment has been carried out the man in the street may not have an adequate opportunity to question the validity of the assessment.

Local planning authorities will shoulder a heavy burden to ensure proper access to information contained in any environmental impact assessment. They will need to allow proper opportunities for public comment and objections. If local planning authorities are to be seen as being fair and helpful to the lay public they will have little to fear about criticism on this score, although the government may find itself under mounting criticism because of the problems that the individual will inevitably find for the reasons already given.

The Use of EIA in Development Planning

It is worth considering the position of local planning authorities in relation to environmental impact assessments in their position as statutory planning authorities. Some authorities have good experience of receiving such assessments, particularly those who have experience of oil and gas exploration and those facing problems associated with large mineral extraction sites. Their experience has taught them the value of using an environmental impact assessment in analysing an application for development and for there to be early and full discussion with the developer. As a result of these experiences how will they affect their future working? It may be that some county authorities will consider incorporating policies within their Structure Plan Reviews requiring environmental impact assessments normally to accompany certain types of major development. This would make it clear to prospective developers that certain specified developments have been identified as being of sufficient importance and impact to require such an assessment to accompany any planning application. It is likely that the types of development specified will be outside those specified in Annex I of the Directive. In development control, local planning authorities are likely to specify those cases where an environmental impact assessment will be required in their development control policy guidance notes and here again the cases specified are likely to be outside the provisions of Annex I of the Directive.

One problem which a local planning authority faces when it receives an environmental impact assessment is this, 'is it completely objective and free from prejudice in favour of the development proposed?' This means that the authority must be in a position to analyse the assessment and decide whether or not it is biased in any way. In most cases the authority will not have the necessary expertise to analyse all aspects of the EIA and will have to turn to other agencies for assistance. It is also likely to need the support of its own independent consultant to advise on the partiality of the submitted assessment. For the future, in respect of those major proposals where an assessment is submitted, either under the terms of the Directive or voluntarily, the use of consultant advice to local planning authorities seems inevitable.

Fortunately there seems to be adequate specialist advice available, although some local authorities are concerned that those specialists available to them are likely to be the same persons who are also available to the developers. Present experience in this country has shown that there is independent expert opinion available and it is not anticipated any difficulties should arise because of the independence of the advice likely to be available. It does seem that a new expertise will emerge, that of being a specialist in EC Directive work concerning environmental impact assessment.

Local planning authorities are likely to seek a full analysis of the assessments submitted to them not least because of the pressures which will be placed upon them by their local electorate but also to carry out their statutory functions as local planning authority, and if this means employing consultants it is considered that authorities will do so. Past experience has shown this has been the practice and there is no reason to suppose the position will change in the future.

The authorities will realize the important weight which has been placed on their shoulders in respect of those few Annex I cases that come their way and are likely to take the responsibility of adjudicating on such proposals in a most responsible way, endeavouring to strike the balance between the proposals of the developer and the needs of the community.

Speed and Delays

The government has suggested that the period for considering a planning application which is accompanied by an environmental impact assessment should be sixteen weeks instead of the normal eight-week period. It has been argued that this is too short having regard to the impact of such a major development and the need for full and proper public consultations. If the assessment which accompanies the planning application has been produced after full discussion with the appropriate authorities it should be in such a form as to allow public participation to start from the day of the submission of the planning application, in which case the argument of the government that sixteen weeks is not an unreasonable period for all the consultations and public participation to take place seems acceptable. But again experience will show whether this period of sixteen weeks is adequate or not.

The period of sixteen weeks in which the local planning authority has to consider a major application such as those falling within Annex I of the Directive cannot be considered to be unreasonable from a developer's point of view and hardly likely to be a delay factor in a project of such a magnitude with an obvious long lead time from planning to implementation. The extension of the period of time allowed to local planning authorities to consider a planning application, with its consequent possibility for delay, could act as an inducement to the developer to ensure he had full and early discussions with the relevant authorities.

It could be argued that by allowing a local planning authority a longer period of time to determine this class of application falling with Annex I of the Directive it may obviate the necessity of 'call in' by the Secretary of State. On the other hand if the Secretary of State does wish to 'call in' the application there may be an inducement on him to do so early within the sixteen-week period. If the planning application is rejected by the local planning authority at the end of the sixteen-week period the developer will only be worse off by two months compared with an ordinary application for which a period of eight weeks is allowed for decision, but as had been said earlier in a project of the size set out in Annex I, it would normally have such a long lead time and this would be of no serious consequence.

The government expects that the production of an environmental impact assessment will assist local planning authorities to come to a decision and will be looking for a reduction in appeals in respect of major developments as a result. This may be a laudable expectation but local planning authorities when considering these major proposals will be ever watchful of local interests. It does not seem therefore that an environmental impact assessment of its own volition will reduce the prospect of planning appeals – it ought to make the decision-making process for the local planning authority that much easier, and in doing so ensuring local interests are fully aware of the implications of the proposal.

The Department of Environment Consultation Documents

The introduction of statutory environmental impact assessments must be welcomed, and should help to ensure better decisions are reached on planning proposals which raise major environmental issues. The attitude of the govenment towards Annex II projects is unfortunate, and because it is so restrictive the full benefits of the Directive are unlikely to be achieved. The number of developments likely to fall within Annex I are very small indeed and proposals of great significance are likely to have environmental consequences which fall outside this Annex and will not be subject to any statutory assessment. The Royal Town Planning Institute expressed its disappointment at the exclusion of some of the categories in Annex II from statutory EIA requirements and argued that environmental impact assessment should cover a much wider range of topics than those set out in Annex I (Royal Town Planning Institute, 1986). There are many others disappointed at the limited approach of the government. Their failure to take on board other categories of development may well not comply with the spirit and letter of the Directive.

Central government should also recognize the number of voluntary EIA's presently submitted, many of excellent quality, and should encourage their presentation rather than discourage it, which appears to be the view of the Department of Environment. The Department's view on the content of environmental impact assessment is also unhelpful, particularly where it refers to the need for original scientific research being required only in very exceptional circumstances. The government proposals have sought to strike a balance between the need for an EIA in relation to a major development and to restrict placing a burden on the developer by requiring such a detailed study. Some will consider a satisfactory balance has been struck. Others will consider the government has been less than even-handed and has favoured the developer. The draft Advisory Booklet (EC Working Party, 1985) was found to be helpful by the Royal Town Planning Institute (1986) which suggested it would be inappropriate for the local planning authority to become jointly involved with the developer in providing information about the proposed development. It wanted the local planning authority to be seen as totally independent from the developer and thus able to carry out its statutory duty in an unfettered manner.

Local planning authorities will face difficulties if there has not been early and full pre-submission discussions, since they will only have twenty-eight days in which to decide whether the EIA is acceptable. This may be insufficient time and may lead to a refusal to accept the assessment in its submitted form, thus leading to argument and delay, the very object the procedure needs to overcome. Local authorities have an opportunity to influence the preparation of the environmental impact assessment through the provision of information and this is an opportunity that must be readily grasped. By making information available to the developer, the local planning authority is also likely to make the assessment more understandable to local inhabitants who will recognize locally gathered information and be able more readily to understand it. The Royal Town Planning Institute was disappointed that Ministers had rejected the proposals of the Working Party with regard to the pre-application procedures which it considered would have been beneficial to developers, planning authorities and the public and would also have ensured a better opportunity for the preparation and presentation of the environmental impact assessment (Royal Town Planning Institute, 1986).

The Institute was also critical that the Consultation Papers appeared to pay little regard to the need to show in the EIA that alternative locations had been examined and felt that the principal conclusion of the Working Party that the requirements of the Directive could be met without imposing significant new burdens on planning authorities was wrong.

The Wider Implications

It may be considered ironic that the government is introducing the requirement of environmental impact assessments at a time when it is heavily engaged in deregulation, particularly in the town and country planning service. However, the way in which it is intended that the environmental impact assessment system will be introduced with possibly not more than twenty cases per year, is unlikely to cause problems and, in any case, the Department of the Environment Working Party concluded that the requirements of the Directive could be met without imposing significant new burdens on either developers or planning authorities. No doubt in coming to this conclusion the Working Party was well aware of those proposals for development which were presently accompanied by an environmental assessment and such cases fell outside the provisions of the Directive. The considerable limitations of the Directive are such that they are unlikely to adversely affect the present deregulation programme of the government, nor will it in any way prejudice it.

There is little doubt that at the present time the government is recognizing the strength of the conservation movement and its wish to ensure that adequate protection is being given to the environment of the country. For this reason it will feel that by the introduction of the provisions of the Directive it is responding to these pressures, but will conservationists feel that the Directive goes far enough? It would seem that once it is appreciated how restrictive the Directive is in operation there will be strong pressure to bring in most of Annex II projects. There is evidence that where environmental impact assessments have been produced a better understanding of the problems associated with the proposed development have occurred (Carless Exploration Ltd, 1984; British Petroleum Ltd, 1986), and there is likely to be a continued demand for such assessments in the future. If as a result of the limited nature of the Directive this does not happen it seems likely there will be strong reaction from many amenity bodies and national societies for the increased use of environmental impact assessments which will put the government under pressure to review its position with regard to Annex II projects.

There are many practitioners and environmentalists who recognize the strengths of the environmental impact assessment system and believe, despite its disadvantages, these are clearly outweighed by the advantages.

Yet, the claimed value of the Directive will only be achieved if environmental impact assessments produce a better understanding of major proposals, a better chance for local democracy, a lessening of outright opposition to proposals because of lack of information and the opportunity for better decision making because of a full and proper understanding of the proposed development.

NOTE

1. EC Environmental Assssment Directive Working Party is a working party set up by the Department of the Environment consisting of representatives from government departments, local authority associations, amenity bodies, industry and the Royal Town Planning Institute.

REFERENCES

British Petroleum Development Ltd (1986) Planning application for extraction of oil and gas at Wytch Farm, Dorset and at Fursey Island, Dorset.

Carless Exploration Ltd (1984) Planning application for extraction of oil and gas at Humbly Grove, Hampshire.

Commission of the European Communities (1985) Council Directive 85/337 of 27 June, 1985, on the assessment of the effects of certain public and private projects on the environments. *Official Journal of the European Communities*, No. L 175/40 5.7.85. Brussels.

EC Environmental Assessment Directive Working Party (1985) Draft Advisory Booklet; *Environmental Assessments and the Planning of Major Projects*. London: Department of Environment.

Royal Town Planning Institute (1986) Observations on the Department of the Environment Consultation Paper: Implementation of the European Directive on Environment Assessment. London: RTPI.

Chapter Ten

Total Assessment: Myth or Reality?

DAVID COPE and PETER HILLS

The 1969 US National Environmental Policy Act (NEPA) and its associated system of Environmental Impact Statements was a political response to over ten years of intense debate on environmental issues. Required to elicit this response, as Ashby (1978) has cogently demonstrated, were both the systematic academic analysis of environmental processes and also a few spectacular incidents of environmental pollution which, through media attention, thrust the environment to the top of the the the political agenda (Solesbury, 1976).

It is sometimes argued that concern with environmental issues has subsequently been eclipsed by more pressing contemporary problems such as maintenance of employment and energy supply difficulties. In reality, the strident and frequently ill-informed excitability which surrounded early 'scientific' and political forays into environmental subjects has been toned down, so that a 'second generation' level of environmental research and policy activity – less spectacular but more methodical – characterised the field by the end of the 1970s. Meanwhile, academic research in areas of assessment[1] activity which relate well to the issues occasioned by, for example, employment and energy considerations, is steadily growing in scope. Events in such fields tend to be more incremental and less spectacular in form than those which precipitated the political response to environmental problems. However, it is reasonable to expect increased attention to be devoted to assessment procedures which lend themselves to handling non-environmental factors or which subsume environmental considerations within a wider evaluation remit and it is to these that this paper is addressed.

For Britain and the other Western European countries, US experience since the creation of the Council on Environmental Quality (CEQ) and the Environmental Protection Agency (EPA) has provided a valuable test-piece which can be judged selectively for components relevant in a European context and those which are appropriate only within their country of evolution (Clark et al., 1978; Garner, 1979). The entire current discussion of environmental impact assessment (EIA) within Britain and Europe is a specific example of a general process of diffusion of social, political and economic institutions and procedures outwards from the United States to the wider world, a trend which has been described generally by writers such as Rose (1974) and Davie (1972). This is not to maintain that the United States had a monopoly of initiative in the development of assessment procedures (cf. the Swedish Energy Commission study (1978) in the field of risk assessment), but simply to argue that examination of the trend toward 'total' assessment which has characterized academic and political developments in the United States may have more than theoretical interest for British and European observers. Wandesford-Smith (1979) has provided a succinct analysis of the varieties and application of EIA procedures in different countries.

The emergence and evolution of EIA to evaluate processes affecting the physical environmental and to inform responses to them has, therefore, been paralleled by, and in some cases has actively stimulated, work in other fields of concern where procedures have also been sought for assessing the consequences of change. The loose term 'fields of concern' is used deliberately here for a common characteristic of much of this work is that it has occurred in areas which straddle traditional disciplinary boundaries and has encouraged new foci of research effort, some of which may be evolving into disciplines in their own right. The effort has already produced some institutional response, of which the US Office of Technology Assessment (OTA) is the prime example.

A prominent feature of these several assessment procedures is the presence of many issues of common concern in their theoretical underpinnings, methodologies and substantive attention. These considerations alone would provide the basis for interesting comparisons but beyond this there seems to be a strong centralizing process occurring with pressures tending to draw various assessment dimensions together. As the title of one paper on this matter – 'The Complete Assessor' – suggests (Walker and Black, 1978), it is possible to envisage the emergence of a 'comprehensive' or 'total' assessment procedure. This would incorporate an evaluation of *all* facets of a particular development proposal, or a general policy, and would set physical environmental impacts alongside technological, social, economic, political, epidemiological and other factors. A widespread conviction exists that a critical need is to evolve the capacity to address the inter-relationships *between* these various thematic dimensions. This mandates some form of unitary comparison procedure.[2]

Trends in EIA

The evolution of EIA has, to an extent, followed in a compressed form some of the trends in the evolution of geography itself. There has been a progressive move from an exclusive concern with physical environmental impact – air and water pollution, physical land-take, implications for flora and fauna, etc., – towards a wider concern embracing the social and economic impacts to which a development proposal will give rise. This trend has been encouraged by the recognition that the total physical environmental impacts of a proposal will be heavily conditioned by the higher-order social, economic and political circumstances which are attendant upon it, such as the location policy adopted for housing of employees, the general employment multiplier effects and so on. In some situations the physical environmental effects of such associated features may outweigh those which inhere to the development itself.

Beyond feedback relationships on the physical environment, which result from the social and economic characteristics of a particular development proposal, there is also a recognition that these consequences must be evaluated alongside those which affect the physical environment, if any assessment procedure is to be analytically coherent and politically acceptable. In EIA the definitional interpretation placed on the word 'environment' has been successively expanded, not only to include the 'built' as well as the natural environment but also, increasingly, conceptions of the 'social' and 'economic' environment. Stoel and Scherr (1978), commenting on the development of EIA, note that in the US situation,

> the general rule adopted by the courts has been that if an action has significant consequences for the physical environment, the impact statement must address social consequences as well . . . however, . . . if an action has only social or economic impacts, no statement is required.

The clear implication is that consideration of social and economic impacts is contingent upon, and secondary to, those of a physical environmental nature and this may be at least a partial explanation of the emergence of 'social impact assessment' as a distinct area of inquiry in the United States. In Britain, a holistic conception of 'environment' has always had a wide acceptance. Catlow and Thirlwall (1977) noted the range of interpretations which can be placed upon the word and opted for one which considered

> the whole picture of the human environment in the area affected by a proposed development (including) man's social and economic well-being as well as his enjoyment of the physical and natural environment in which he lives and works.

Putting aside the ethical question of whether physical and natural environmental considerations should be restricted to those which lead to human 'enjoyment', or whether the physical and natural environments have 'standing' in their own right (Tribe et al., 1976), such a definition is clearly extremely embracing in scope and was even seen as including 'disturbance to the concept of man in the perspective of history'. Similarly, in its 1979 report, the Watt Committee on Energy noted that for its purposes the term 'environmental impact', will not be as wholly comprehensive as it can be constructed. It is intended to embrace the physical, economic and social consequences of development of energy resources, conversion, storage and distribution

It is difficult to envisage the scope of the even wider construction of environment which the authors had in mind, unless this was the use of the term sometimes found in economic and management literature where it refers to all the circumstances external to a particular business or agency which condition that entity's operations.

Social Impact Assessment

While the scope of EIA has expanded there has also arisen, as noted, a separately identifiable area of assessment activity which has come to be known as social impact assessment.[3] One spur to this development has been a reaction against the dominance of physical environmental considerations in project assessment, particularly in the United States (Boothroyd, 1978). Thus, Vlachos *et al.*, (1975), commenting on the progress of EIA since the 1969 NEP Act, regretted that in its operation 'distinctively social impacts have tended to be implicit, indirect and qualitative'. Social scientists have been at pains to establish the legitimacy of their concerns through their incorporation into assessment procedures carrying the status of legislative requirement. As Stoel and Scherr (1978) note, it has been the fact the EIA is enforceable by litigation which has spurred federal and other agencies to accord environmental factors the attention that they have received in recent years. Another encouragement to the development of social impact assessment has been the emergence in many specific development proposal assessments of a fundamental conflict between employment and environmental considerations. 'Jobs vs. the environment' has become a recurrent dilemma, and the agencies responsible for advancing the causes of these conflicting interests have frequently retreated to stereotyped positions with limited possibilities for flexible appraisal of particular circumstances. Social impact assessment has been seen as a procedure to ensure the accordance of equal significance, in project and programme appraisal, to non-environmental (and possibly less directly quantifiable) but politically important factors. The implication in much social science discussion of NEPA has been that it resulted in an over emphasis on physical environmental factors. Vlachos *et al.* (1975) say of the US situation that 'recent environmental legislation has made it imperative to study the integration of social sciences with natural sciences, the estimation of social costs, the effects of public projects on communities and the social policy alternatives concerning various technological changes'. Boothroyd (1978) notes that within SIA itself

it appears that the trend is toward a broadening of the things considered by SIA, from individuated, purely physical developments . . . to more abstract items such as community growth, technological innovation and even such causal items as immigration policies.

Aside from this balancing function, there are other attitudes and insights which it is asserted SIA has brought to the assessment process. Certainly it is true that sociological interpretation will tend to ensure that the conflict dimensions of any particular situation are not disguised in a synthetic overall 'felicific calculus' and that due recognition is given to the fact that equivalent costs and benefits of any policy or project are rarely received by the same interest groups. Indeed the entire discussion of 'equity' issues in assessment procedures has received a major stimulus from the explicit pursuit of a social and political dimension as input to such exercises.

Another perspective which it is claimed stems from the emergence of SIA is the recognition of the limitations of uni-directional models of causality which tend to dominate physical environmental considerations. The strength of SIA is its firm awareness of mutual causality – the existence of complex strands of feedback and feedforward relationships which invalidate any simplistic analyses of cause/consequence chains (Vlachos *et al.*, 1975). An operational limitation of SIA has been the tendency of many practitioners to attempt to bolster its credibility *vis à vis* EIA by concentrating on those aspects of the social environment which lend themselves well to quantification, such as demographic variables and pattens of geographical distribution of population, while relatively neglecting other less readily quantifiable but not necessarily less significant social factors.

Risk Assessment

A singular feature of many major industrial and other developments initiated or proposed in the past has been that they have involved activities potentially hazardous to human life or health (Health and Safety Executive, 1976). The prime example (at least in public awareness) has been nuclear power, but the development of road and air transport facilities, location of petrochemical and other material processing plants, storage facilties, mining, quarrying and hydro-electric schemes all have had implications in this field. Of course, much of the concern which has led to the development of EIA has stemmed from the existence of environmental pollution pathways resulting in hazards to human life and health, and the embracing definition of environment adopted by Catlow and Thirlwall (1977) includes as one of its impact groups 'impact on health, safety and convenience', including pollution, waste disposal, noise and traffic accident hazards. The political significance of health and safety aspects of developments and activities are such that they are invariably key factors in the operational and locational decision-taking process applied to any proposal. A number of catastrophic events such as Flixborough, Three Mile Island, Bophal, and Chernobyl have served to heighten the emphasis placed on assessment of potentially hazardous installations and activities and this matter has increasingly become one for which planners at the district and county levels have needed to develop coherent and mutually consistent policies (Hankey and Probert, 1978).

The requirement for assessment procedures which can analyse the risk levels involved in various activities of industrial society, and thereby determine the levels of design and the operational and location safe-guards required to insure against accidents of every scale, led to the emergence of Risk Assessment as a very active field of concern in the United States and to a lesser extent in Western Europe (Starr *et al.*, 1976). The influence of nuclear power in stimulating the development of this research effort has been critical. The report of the US Nuclear Regulatory Commission (the Rasmussen report) (Rasmussen, 1974) and the Swedish Barsebäck II study (Statens Kärnkrafts Inspektion, 1978) are two examples (with somewhat conflicting results) of major studies of the risks involved in routine operation of, and accidental circumstances at, nuclear installations. A report by Inhaber (1978) of the Canadian Atomic Energy Authority attempted to compare the risks involved in nuclear power generation with those from a range of other sources of energy. Although some of the findings of this study have subsequently been heavily criticized, it does demonstrate through its methodology an important characteristic of all risk assessment procedures – the rigorous pursuit of networks of causal linkages in the hope that all potentially relevant circumstances are taken into account. In the context of nuclear power, this is to ensure that every conceivable combination of accident-generating circumstances is recognized, its probability computed and its consequences assessed. It was this rigorous pursuit of chains of circumstances which led the Inhaber report to its somewhat surprising and subsequently disputed finding that some 'alternative' energy supply options were, when the risks associated with the production of their requisite infrastructures and their operational maintenance were assessed, more dangerous than some nuclear power options.

Risk assessment has had to develop methodologies which are capable of handling problems which lie in the field of 'trans-science' (Weinberg, 1972; Council for Science and Society, 1977). Experimental research on such problems is either too expensive, too dangerous or requires too long a time period in which to produce results. In addressing the attention of scientists and engineers to such problems, risk assessment has drawn them closer to the non-experimental fields of analysis based on association and extrapolation which are the familiar territory of social and environmental researchers. The role of risk assessment in bringing together researchers from a wide range of academic disciplines so as to focus their expertise on concerns of considerable significance is probably a major factor explaining the popularity which it is currently enjoying as an area of policy research. The amount of comment which followed publication of the Health and Safety Executive's (1978) study of potential hazards from operations in the Canvey Island area, one of the few UK examples of an attempt at a wide-ranging risk assessment exercise, suggests that recognition of the significance and attraction of the approach is beginning to emerge in this country. However, it is indicative of current capabilities in this field that the Executive should have to commission outside consultants to produce the study and that the Health and Safety Commission (1978) should see the study as a 'unique request'.

Until recently, risk assessment tended to concentrate heavily on risks associated with mortality and morbidity. Although, as mentioned, these areas present considerable problems of experimentation and therefore of quantification, they do offer some opportunities for 'hard' analysis in that reasonable sets of associational data are generally available on which to base estimates of risk. Assessment of less corporeal risks, such as the decreased incidence of civil liberties which might stem from the emergence of the 'plutonium economy' (*Justice*, 1978) or the international political problems attendant on nuclear proliferation (Department of the Environment, 1978; Royal Commission, 1976; Town and Country Planning Association, 1978) has figured in few studies. Even socio-economic risks such as the levels of employment security associated with different development options have not received much methodological or substantive attention, although the risk assessment panel of the US National Academy of Science's Committee on Nuclear and Alternative Energy Systems (CONAES) study has attempted to include such non-epidemiological risks in its evaluation procedures (CONAES, 1979).

Probably the most intractable and, at the same time, fascinating topic within the field of risk assessment is the recognition and evaluation of 'objective' and 'perceived' levels of risk (Sjöberg, 1977). As many studies have shown, public awareness and acceptability of various types of risk are not linear. Generally speaking, risk levels which result in the frequent occurrence of a small number of fatalities tend to attract a lesser amount of risk-aversive behaviour than infrequent events with larger impacts, although the sum totals of fatalities from such different causes may be comparable. The distinction between what are termed 'individual' and 'societal' risks in the analysis carried out by the Health and Safety Executive (1978) on Canvey Island is founded on this non-linear perception of risk impacts. The study team used an arbitrary definition of 'societal' impacts as involving the potential deaths of more than ten persons. Sociological and psychological analysis of hazard and risk perception and the ways in which trade-offs are established between risks and associated benefits are likely to be of increasing importance in the development of risk assessment (Lowrance, 1976).

Technology Assessment

In recent years there have been so many examples of the fact that newly deployed technological developments may have significant but wholly unforeseen side-effects that drawing attention to this may seem somewhat redundant. Sometimes side-effects have been beneficial (the most famous and possibly over-rated being the development of 'non-stick' cooking utensils from materials used for coating the fuel tanks of space rockets) but more often than not impacts have been deleterious physically, physiologically, psychologically or socially. In some specific instances, especially in the fields of pharmaceuticals and biochemical agents, the side-effects have been disastrous. Confronted with such experiences, it is not surprising that a strong conviction should arise that there should be some means, and the agencies to mobilize such means, to examine incipient technological developments and their potential consequences to provide forewarnings of impacts, and if necessary to interdict or modify the introduction and diffusion of the technological development. Such activity has come to be known as Technology Assessment (TA) and is undoubtedly the most far-reaching and comprehensive of the different assessment approaches discussed in this paper. It would be wrong to imagine that TA represents a strict chronological culmination of the development of assessment procedures which began with the postulation and codification of EIA. As Medford (1973) has documented, political action to establish some form of procedure for technology assessment in the United States dates back to at least 1967 and the OECD was advocating independent 'look-out' agencies to assess 'technological initiatives and developments' as early as 1971.

TA has been defined as

the systematic study of effects on society that may occur when a technology is introduced, extended or modified, with special emphasis on the impacts that are unintended, indirect and delayed . . . technology assessment is a policy generation and analysis or decision-oriented tool. Its aims, techniques and methods are best conceived as yielding systematic inputs into the larger political economic decision process. (Coates, 1974)

TA has therefore been conceived of as an activity which could reduce the number of unforeseen side-effects that flow from a particular technology; which could lend itself to analysis and evaluation of long-term issues; which could be an effective tool for producing policy-forming advice for decision-takers; and also which could be a mechanism for stimulating wider public awareness of pertinent issues in the field of technology impacts. It was the declared hope of the initiator of the legislation creating OTA in 1972, Congressman E. Daddario (subsequently the first director of OTA) that 'knowledge broadly disseminated is the key to reducing uncertainty in our society'. Uncertainty, particularly over environmental matters, was held to have resulted in certain problems being treated with more urgent priority than they objectively warranted, while

legislative or regulatory action [may be] rushed prior to the development of sufficient information . . . [and] the resultant governmental actions may not necessarily be in the best interests of the public they are meant to protect. (Daddario, 1977)

However, as Daddario (1977) specifically acknowledges,

> environmental concerns were prominent among the issues which came into focus
> during the 1960s that led some members of Congress to recognize the need for an
> enhanced ability to anticipate the impacts of technology, both beneficial and
> adverse.

It is probably true to say that the focusing of public attention on
environmental matters and the passing of the 1969 National Environmental
Policy Act were necessary precursors for the development of TA techniques and
certainly for its political reification in United States through the Technology
Assessment Act of 1972 and the subsequent creation of OTA in 1974. Thus, in
the declaration and purpose section of the Act it is noted that technological
impacts are

> increasingly extensive, pervasive and critical in their impact, beneficial and
> adverse, on the *natural and social environment*. (Technology Assessment Act,
> Public Law 92–484, 1972, emphasis added)

The subsequent history of OTA and, by extension, of TA itself has been
somewhat chequered and to understand this it is necessary to appreciate the
political context of the creation and activities of the Office. Unlike NEPA and its
associated system of EIS, OTA was created as an advisory area of Congress
designed, at least partially, to rectify the low level of appreciation of, and
competence to review, scientific and technological matters on the part of
Congresspersons. Such deficiencies are not restricted to the US political system,
nor were levels of competence significantly higher for other topic areas such as
environmental issues, but a Congressional review function was not part of the
NEPA legislation.

In trying to bridge the interpretational gap between legislators and the science
and technology community, OTA has sometimes fallen victim to the difficulties
inherent in such a role. At first the office maintained closer contacts with the
political end of the bipolarity. This meant that political dispute sometimes
infiltrated what were supposedly the objective, advice-providing functions of
the agency. One particular area of tension has been the question of short-term
vs.long-term foci of concern, with the political community naturally
gravitating towards the former but the science and technology community's
sympathies lying more naturally towards the latter. The question of the activity
emphasis of the Office has tended to focus on the independence of the director
and the role of the twelve-member governors' board drawn from the Senate and
House of Representatives. The specific investigations which OTA undertakes
arise from requests channelled by Congress Committee chairpersons, from
suggestions made by the governing board or by initiative of the director 'in
consultation with the board'. Clearly the greater the freedom of initiative of the
director, which has tended to grow in recent years, the less the influence of the
Congressional board and the greater their, and by extension, Congress' sense of
being passively advised rather than actively initiating assessment activity

(Dickson, 1979). The tensions to which this dichotomy has given rise have been strong enough to show up even in as bland a document as the Annual Report of the Office (OTA, 1978). Another characteristic of the operations of OTA which might be expected to lead to various tensions is its assessment of the programmes of government agencies, such as an examination of the Internal Revenue Service's proposed tax administration system because it was 'on the leading edge of the state of computer art'; the proposed National Energy plan; and the operation of the EPA from 1976 to 1980. Such activities are inevitably required for effective evaluation of impacts of technologies and their supportive legislation but the difficulties which arise from the reluctance of government departments to submit fully to open inspection of their activities by the legislative can be easily appreciated.

As a newly emerging and embracing area of assessment activity, TA does not have a rigorous and generally accepted corpus of theoretical underpinnings and methodology. Practice has been to draw in an eclectic fashion on techniques developed in other disciplines. Indeed, as one advocate has described it, 'technology assessment is an art form . . . It is separate, distinct and different . . . and . . . implies both a degree of creativity as well as the use of techniques' (Coates, 1973). It is however possible to identify some features, characterizing most TA exercises, which derive from the general aims of the activity.

A survey of current circumstances of relevance to the topic under examination is a necessary starting point. Virtually as essential is some form of forecasting activity. The long-term focus of TA has been emphasized in the definitional description outlined above. Given that TA has as an axiom the consideration of alternatives to the technology under scrutiny and the various contexts in which the technology may operate, such forecasting tends to be of the conditional 'if A then X' type, usually based on scenario construction. Recognition of the sociological dimension also tends to be implicit in TA. Most studies look at the various interest groups affected by the technology under consideration and their relative powers in influencing the technology's deployment and consequences. The socio-political dimensions tend to be introduced to studies after an initial phase which has concentrated on technological considerations, probably because to introduce components of conflict at the beginning of a study might prevent it from taking off at all.

TA carried out on geothermal energy under the national Science Foundation's 'Research Applied to National Needs' programme provides a good illustration of a TA exercise (Futures Group, 1975). The study began with a 'state of the art' exercise based on conventional reviews of literature sources. The currently existing inter-relationships between geothermal energy and other factors, technological, economic and social, were elucidated with the aim of identifying those which were potentially significant as influences on, or consequences of, the technology under scrutiny. Forecasts were made by a variety of means, including simulation modelling, questioning of a panel of experts and the application of 'Trend Impact Analysis', a modification of extrapolatory forecasting which 'permits extrapolations of historical trends to be modified in view of expectations about future events which could influence the trends). Analysis of policies which might affect the technology under scrutiny included identification of agencies which could initiate impacting policies, costs of policies, intended and unintended policy consequences and effects of the timing of introduction of the policy. Another feature of the study which, given the nature of the issues to which TA is addressed, tends to feature fairly frequently in TA exercises was the evaluation of two distinct evolutionary pathways – a 'normal' and a 'crash' development programme. Discussion of, if not the inevitability of, 'crash development' seems to be the inescapable corollary of policies for 'keeping options open' which characterize the energy, and some other, policies of so many Western governments.

A Critique of Assessment Concepts and Procedures

It is as well to note that circumscribed interpretation on notions of 'totality' or 'comprehensiveness' is applied to assessment. Such qualities can only exist within the limitations of what Tribe (1973) has called 'instrumental rationality'. Tribe has provided a masterly philosophical critique of the entire concept of 'rational' assessment which alongside the more strident critical overview of Lovins (1977) points out both imperfections in assessment procedures as conceived and carried out to date and also some intrinsic limitations of the approach which are probably insuperable. Defining instrumental rationality as 'selection of efficacious means to previously given ends', Tribe notes that there has been an inexorable trend on the part of institutions to 'extend prevalent notions of rational planning and policy evaluation so as to broaden the range of factors reflected in technological choice'. The implication that the broadening of assessment procedures has been a conscious, pre-determined policy on the part of various establishment institutions overlooks the extent to which these institutions have been pushed into incorporation of more and more dimensions by the criticisms levelled at their existing decision-taking support systems. Lovins (1977) similarly sees in assessment procedures an attempt 'to grease the tracks with rationality in order to avoid emotional impediments to technological advance'. However, at least in Britain, the demand for wider, 'more relevant', assessment has often come from those whose aims have been to put boulders on the technological tracks rather than to grease them.[4] Many development agencies view with despair the ability of opposition groups to raise a plethora of issues of infinitesimal relevance which nonetheless they may be obliged to check out through exhaustive analysis.

Whatever the origins of the ever-increasing demands for assessment and the ever-widening notions of which topics are relevant to incorporate in assessment procedures, it remains true that the assessment process as such is founded on conceptual principles which do not embrace all the components of human action and motivation. In practice to date they have certainly been conducted with procedural 'blinkers'. Of these, Tribe (1973) notes the general tendency to reduce the various dimensions of a complex problem to such an extent that the conceptual model used in the assessment process tends to mis-represent the real-world problem. In particular, the 'soft' variables tend to get squeezed out to the extent that 'entire problems' [are] reduced to terms that mis-state their underlying structure and ignore the 'global' features which give them their total character'. As Tribe (1973) notes, this is not an endemic weakness of the concept of assessment through instrumental rationality and, as a broad generalization demonstrated by the discussion earlier in this chapter, the trend has been for a successively greater degree of incorporation of such variables.

Less progress has been made towards eliminating the second of Tribe's criticisms of assessment practice – the tendency to focus on end results rather than process, in particular the range of possible processes, for achieving the end results. Tribe notes that processes nearly always have an ascribed value in their own right and, especially in the field of technology, this may transcend consideration of end results. Development and deployment of a technology becomes an end in itself as many have argued has been the case with aeronautical and nuclear engineering. The processes of interaction between a technology and society may have more significance in societal terms than the actual substantive outcomes of the technology and this fact needs to be accorded due weight in any assessment procedure which lays claim to being comprehensive.

The major intrinsic weakness of assessment based on instrumental rationality is its inability to handle the normative considerations which attach to goals. Assessment procedures do not extend to the questioning of the value systems on which decisions are based and do not lend themselves to establishing what these value systems should be. Decisions on technologies alter the circumstances of the population on which they have their impact and in so doing the preferences and consciousness of the population are altered over time so that the characteristics of the affected population as social groups and as individuals are changed. As Tribe (1973) notes, 'most of the decisions which will be remembered as landmarks in technological and environmental history a century hence (will be those whose) dominant dimension will probably be their impact in reconstituting who we are.' Instrumental analysis of these decisions alone is unlikely to be indicative of the appropriate paths to follow. Tribe admits the value of assessment as an 'early-warning' activity to highlight second and subsequent order effects of a technology and argues that procedures based on instrumental rationality should be set alongside, and thereby enriched by, assessment of technologies in constitutive, normative terms as well. The following discussion of 'total' assessment is developed with concurrence to Tribe's exegesis of the limitations of the assessment activity.

Lovins also accepts the value of assessment procedures. He identifies: the extraction of the fullest information from development-initiating agencies, the overt presentation of the factors held by actors to justify or repudiate a development; and 'rigor in approaching the genuine limits of quantification', but describes these as 'ancillary' and needing to be set against (through some form of assessment of assessment?) the dangers to which techniques may also give rise such as the discrediting of valid methodologies by their being associated with unsupportable claims for the assessment process as a whole and the risk that politicians confronted with assessment procedures may fail to discharge their political responsibilities either by hiding behind or being substituted by the procedures.

Some Necessary Attributes of 'Total' Assessment

The assessment procedures discussed in this chapter are clearly not distinctly separate entities addressed to the resolution of policy issues in distinctly separate areas. Rather they are interwoven into a complex set of conceptual, methodological, organizational and consequential overlaps. One approach may lend itself well to assessment of individual development proposals while another may be better suited to assessment of broad-scale policies and programmes.[5] One may 'enjoy' a higher degree of legitimacy through legislative back-up; another may depend on its intrinsic academic appeal or possibly on public pressure for its application. Some approaches are appropriate to 'one-off' exercises; others fit better the requirements of continous policy review.

Any system of 'total' assessment operating within the limits of analysis described in the previous section would tend to draw together these various strengths and relevances into a coherent, interlinked procedure for recognition and appraisal of significant developments. A central requirement of such an integrated system would be a well-defined and dynamic procedure for assessment which could accommodate both policies and the individual projects which relate to them. The need for such a hierarchical, nested or tiered system of assessment has long been recognized (O'Riordan, 1976) and the development of TA provides an approach which is well suited to broad-scale appraisal to which more specific assessment exercises, such as EIA or TA, might be linked when there arose an individual project stemming from a previously examined general policy. Obviously such an assessment structure would have major components geared to periodic, if not continual, review of assessed subjects, with special attention to the policy implications shown up by individual projects and the project implications of policies though as iterative examination procedure. The various operational problems which attach to the devising of such a hierarchy of assessments have been well discussed in the sources mentioned and will not be considered further here.

To accommodate both broad policy and specific projects demands a wide range of competence. Procedures for handling time-horizons will need to be systematically codified. Short-, medium- and long-term impacts of policies, programmes and projects will have to be clearly identified. The process of unfolding of a policy or programme, or even some aspects of projects, is as significant as end results while the timing of occurence of impacts will undoubtedly influence decision-taking about the impact-generating entity. The niceties of discounting long-term impacts require a separate paper in their own right (Mishan, 1977; Routley, R. and V., 1978) but the handling of long-term impacts will probably be the most intractable temporal difficulty facing any 'total' assessment procedure because dilemmas over the extent to which the interests of 'future persons' should be incorporated in the assessment are close to the limits of instrumental rationality.

Analogous to the requirement for a rigorous time perspective is that for a broad spatial perspective. Policies will be determined, and work themselves out, at the international or national level while individual projects will of course be locationally specific. As higher and higher order impacts are pursued so, inexorably, the spatial 'impact field' will broaden but in particular cases circumstances are such that a rapid escalation to global impacts occurs even when considering an apparently limited option (e.g. the CO_2 issue in the case of fossil fuel energy policy or weapons proliferation in the case of nuclear power). If assessments are to lead to policy initiatives these may need to be taken by international agencies and consequently these agencies will require competent assessment abilities.

Any 'total' assessment procedure must, quite obviously, be capable of identifying and measuring inter-relationships between the entity under scrutiny and external factors while the entity itself will also usually be complex and systemic in nature. As a special case, beneficial and adverse side-effects must be susceptible to recognition. To achieve the greatest utility, any analysis of inter-relationships should have the capacity to compare results from a multiplicity of assessments. Procedures must therefore, at least to an extent, be standardized and replicable rather than individualistic. Recognition of side-effects is a central principle of technology assessment but the extent to which an assessment achieves the minimizing of surprises is probably more of an art than something to be derived from rigorous application of specific methodologies. Some methodologies, such as certain applications of catastrophe theory (Waddington, 1977), do however lend themselves to formal identification of surprise areas and these might be extended to form a focus of research effort.

'Total' assesment necessarily has a considerable requirement for data. Many of the other factors mentioned in this section will have their own consequences in this area. Issues such as data accuracy, the legitimacy of generalizing from data derived from specific cases and, above all, the inherent unknowability of future states will inevitably contribute to the problems facing development of 'total' assessment. In the face of inherent unknowability the proto-discipline of futures studies has already developed some codified procedures, such as the systematic panelling of 'experts' in the so-called 'Delphi' method of forecasting. Assessment procedures will almost inevitably have to make use of expert judgement and will therefore need to be aware of the limitations which attach to such sources of data as well as recognizing the valuational implications of rely-ing on 'experts'. In facing up to uncertainty, great store has come to be laid in current assessment and planning activities on 'robustness', a term borrowed from statistics and referring to the degree of inbuilt resilience of the results of an assessment to variability in the external circumstances influencing the subject of assessment. Closely allied to robustness is 'flexibility' – the ability to remain operationally relevant under variant external conditions. Something of a 'holy grail' in contemporary British planning practice, flexibility may benefit from the more rigorous examination which it would receive as a desired quality in the context of 'total' assessment procedures. A third requirement of this type, which will almost certainly be a sought-after characteristic of 'total' assessment, is the ability to identify procedures for keeping options open (Walters, 1975). Although current national energy policies in most Western countries are based on such a strategy to an almost pathological degree and option preservation may commend itself to indecisive politicians, it clearly has positive instrumental value especially in decision-taking on long-term, complex, highly capital-intensive policy areas.

The accountability of, and degree of public participation in, a 'total' assessment procedure is another topic worthy of a chapter in its own right (O'Riordan, 1976) as is the question of the institutional mechanisms for carrying out assessments. The greater the claim to comprehensiveness made for a procedure, the more there is a need for a participatory component. Consequently the excessive formalism and obscurantist jargon which have characterized too many assessment exercises to date will be inimical to achievement of a comprehensive procedure. Reconciliation of the need to be simple and transparent without slipping into simplistic analysis of what are usually extremely complex subjects will place great demands on the competence of those carrying out 'total' assessments. The needs for 'total' assessment may come to be met through novel forms of participatory institutional arrangements along the lines of the risk advisory service proposed by the Council for Science and Society (1977). The openness of, and degree of participation in, an assessment procedure will probably strongly influence the extent to which it explicitly includes evaluation of social inequities. This element, a salient feature of SIA, is a prerequisite of any approach which lays claim to the title of 'total' assessment. Without it, the desired process of transfer from assessment to susequent decision-taking will be hollow and disappointing.

The indispensable characteristics of 'total' assessment described in this section are by no means an exhaustive listing of desirable attributes. Nor would it be realistic to expect all assessments, directed at a wide, disparate range of assessment subjects, to conform to a single, closely defined, conceptual, methodological and operational stereotype. Flexibility in approach is another necessary criterion and not merely an enforced consequence of the diversity of subject matter. Assessment procedures themselves should be periodically assessed, to ensure their continuing appropriateness and vitality.

A Recent UK Development: the CENE 'Coal Study'

By way of conclusion, and to focus attention on a current specific assessment example, the remit of the 'Coal Study' at present being carried out by the newly-created Commission on Energy and the Environment will be briefly considered. The creation of the Commission, in itself, is evidence of a continuing government commitment to acknowledging the importance of its study area (the recent change of government does not seem to have invalidated this statement), while the formation of an agency to span the concerns of two government departments represents an institutional recognition, albeit small, of the need for consideration of topics lying across the divides of established ministerial responsibilities.

It would be wrong to imply that the Commission's study is the first example of a wide-ranging, long-term assessment exercise. The Sixth Report of the Royal Commission on Environmental Pollution (1976), on nuclear power and the environment, adopted a very broad remit and included a context-setting study of general energy strategy, specifically considered risks and did not duck discussion of the political and even ethical issues which the subject raised. Similarly, the Report of the Advisory Committee on Trunk Road Assessment (Department of Transport, 1977) included a detailed discussion of assessment frameworks as they relate to strategic road programmes and emphasized the requirement to incorporate non-economic as well as economic components. Although, in justifying a somewhat orthodox appraisal methodology, the Committee clearly laid itself open to the critiques developed by Tribe and Lovins, it did acknowledge the need for any assessment to be comprehensive, comprehensible to the public, inexpensive, able to relate to 'decentralized minor decisions' (i.e. projects, in the terminology of earlier sections of this chapter), and 'balanced' in the costs (not necessarily economic) and benefits that it incorporated in its deliberations. Another recognition of the merits of comprehensive assessment can be found in the report of the Stevens Committee on mineral working (Department of the Environment, 1976). Although addressed to a very specific topic, this considered the need for a 'comprehensive statement of the national interest' as it related to mineral workings, within a chapter devoted to long-term planning. In the non-environmental field, mention should perhaps be made of the Central Policy Review Staff (1977) study of the impacts of population trends on the social services. The reference to these studies is not meant to imply that the agencies which conducted them are appropriate models for those which would carry out comprehensive assessment activities.

The 'Coal Study' is not described in terms such as 'comprehensive assessment' or 'technology assessment', etc. in any of the literature produced by the Commission to date, but the terms of reference contain several interesting phrases which demonstrate that the study will be adopting approaches relevant to a broad-ranging assessment (Commission on Energy and the Environment,

1978 and 1979). The 'longer-term environmental implications' of coal are to be examined, 'looking to the period around and beyond the turn of the century'. Singled out for special mention are 'likely new technologies' while the explicit focus on the total coal energy option, from production through distribution, conversion, and end use to residue disposal, represents the holistic approach which is necessary for adequate assessment of broad policies.

The Study's brief specifically excludes 'individual planning cases', so the development of appraisal relevant to particular projects will be in terms of general rather than site-specific considerations. However, the number of coal-related developments in the recent past and present is not so great that the study will be able to cast its analysis in terms of statistical deductions from a large universe of cases. Many conclusions will inevitably be influenced by the exigencies of individual cases. Energy demand and supply forecasting will not feature as part of the study. Instead the intention is to take, as 'the most convenient starting-point', the levels of production envisaged in the Energy Policy Consultative Document (Department of Energy, 1978). However, somewhat obscurely, the Commission has acknowledged that it will need 'to consider the environmental implications of a range of different levels of output' and also production levels beyond 2000 (current long-term plans for the industry have been published only up to the year 2000). By implication therefore, the Commission will concern itself with production scenarios which differ from the stated goals of government and the National Coal Board.

Perhaps the most interesting feature of the study to date has been the apparent widening of the scope of the investigation, in particular a move away from an exclusive concern with physical environmental impacts towards incorporation of social and economic factors. This may have arisen as a result of the thrust of preliminary responses to consultations with potentially interested agencies in which the Commission engaged early in its deliberations. In the latest study outline, under a heading 'Patterns of Environmental Impact', are included sections on 'socio-economic' and employment effects, infrastructure requirements and the locational implications of supply and demand. A separate topic heading covers 'Policy and Planning Procedures' and the most interesting aspect of this section is the intention to identify 'matters that are of common concern to all features of coal production' and which may therefore lead to their being released from the necessity to be discussed at each planning enquiry into forthcoming individual projects. Another stated intention is to 'consider how a more selective and relevant content of local enquiries might be achieved', although the interpretations attached to 'selective' and 'relevant' are not elaborated further.

The evolution of the study deserves to be watched closely as it promises to be a potential landmark in the development of assessment procedures in the United Kingdom. Coal is intrinsically well-suited to be the subject of a pioneering comprehensive assessment. Its physical environmental impacts are paralleled by social, economic, health and safety and political considerations of at least equal magnitude, so that multi-dimensional assessment is a fundamental pre-requisite for any viable assessment. It can be argued that if an assessment approach can be developed to handle the policy implications of the coal energy option, comparable assessments of other options, and indeed studies in most other fields of concern, should be achievable with relative ease. The study is being conducted under the aegis of a Chairman (Lord Flowers), who has already chaired the production of the Royal Commission of Environmental Pollution Sixth Report, itself a singular development in environmental impact assessment activity in the United Kingdom.

Finally, but most importantly, the study is occurring in an intellectual environment of intense interest and considerable activity in the field of assessment procedures in the United States, Europe and the United Kingdom. That this upsurge of interest will fail to influence the study's work is unlikely and, given co-operation and enthusiasm from all parties, the Commission could produce a seminal work in energy, and more general, assessment procedures. If this promise is fulfilled such a study will, alongside the contributions of others such as those mentioned in the introduction to this section, contribute to the United Kingdom's transition through what Tribe (1973) has called the 'fourth discontinuity'–to the harmonious and unified co-existence of man and his technology.

NOTES

1. Throughout this paper the term 'assessment' is taken to refer to the combination of scientific determination of research results (analysis) with the subsequent evaluation of such results for purposes of decision-taking.

2. This procedure would not be confined to, or necessarily consider at all, comparisons in monetary terms and would transcend the recognized limitation of comparison procedures based on cost-benefit analysis.

3. Sometimes also referred to as 'community impact assessment' – an unfortunate term due to the evaluative notions of functional cohesion associated with the word 'community'.

4. Responses to the conducting and outcome of the Thorium Oxide Reprocessing Plant enquiry at Windscale and events at recent road enquiries being the best known examples.

5. For the distinction between policies, programmes, plans and projects, which is very relevant to this section, see Chapter Seven.

REFERENCES

Ashby, E. (1978) *Reconciling Man with the Environment*. Oxford: Oxford University Press.

Boothroyd, P. (1978), Issues in social impact assessment, *Plan Canada*, **18** (2), pp. 118–134.

Catlow, J. and Thirlwall, C. G. (1977) *Environmental Impact Analysis*, DoE Research Report no. 11. London: HMSO.

Central Policy Review Staff (1977) *Population and the Social Services*. London: HMSO.

Clark, M. *et al.* (1978) Environmental impact assessment: an ideology for Europe. *Town and Country Planning*, 46 (9), pp. 395-9.

Coates, J. F. (1973) Institutional and Technical Problems in Risk Analysis. *Proceedings of a Conference on Bulk Transportation of Hazardous Materials by Water in the Future*, Committe on Hazardous Materials, College Park, Maryland, July 9-10.

Coates, J. F. (1974) The Identification and Selection of Candidates and Priorities for Technology Assessment. *Technology Assessment*, February (quoted in Futures Group, 1975).

Commission on Energy and the Environment (1978) *Coal Study: Terms of Reference and Study Outline*. London: HMSO.

Commission on Energy and the Environment (1979) *Coal Study: Organisation and Structure* (CENE 78/21, amended February 1979). London: HMSO.

Committee on Nuclear and Alternative Energy Systems (1979) Final Report. Washington DC: National Academy of Sciences.

Council for Science and Society, (1977) *The Acceptability of Risks*, London: Barry Rose.

Daddario, E. Q. (1977) Uncertainty as Viewed from the Office of Technology Assessment, Keynote address. *Proceedings of the Fourth Symposium on Statistics and the Environment, 3-5 March 1976, Washington D.C.* American Statistical Association, pp. 18-23.

Davie, M. (1972) *In the Future Now*. London: Hamish Hamilton.

Department of Energy (1978) *Energy Policy: A Consultative Document*. London: HMSO.

Department of the Environment (1976) *Planning Control over Mineral Working*, Report of the Committee under the chairmanship of Sir Roger Stevens. London: HMSO.

Department of the Environment (1978) *The Windscale Inquiry*: Report by the Hon. Mr. Justice Parker. London: HMSO.

Department of Transport (1977) *Report of the Advisory Committee on Trunk Road Assessment*. London: HMSO.

Dickson, David *(1979) OTA faces another summer of uncertainty. Nature*, 278, p. 199.

Futures Group (1975) *A Technology Assessment of Geothermal Energy Resource Development*, Executive Summary. Washington, DC: National Science Foundation.

Garner, J. F. (1979) *Environmental Impact Statements in the United States and Britain*, Denman Lecture, 1979, Department of Land Economy, University of Cambridge.

Hankey, B. and Probert, G. (1978) Avoiding industrial disasters. *The Planner*, pp. 139-42.

Health and Safety Executive (1976) *Advisory Committee on Major Hazards*, First Report. London: HMSO.

Health and Safety Executive (1978) *Canvey: an Investigation of Potential Hazards from Operations in the Canvey Island/Thurrock Area*. London: HMSO.

Health and Safety Commission (1978) Covering letter to Secretary of State for Employment accompanying the H. S. E. Canvey study.

Inhaber, H., (1978) *Risk of Energy Production*. Ottawa: Atomic Energy Control Board.

Justice (1978) *Plutonium and Liberty: Some Possible Consequences of Nuclear Reprocessing For an Open Society*, a report based on evidence presented to the Windscale Inquiry. London: Justice.

Lovins, A. (1977) Cost-risk-benefit assessments in energy policy. *George Washington Law Review*, 45 (5), pp. 901-43.

Lowrance, William W. (1976) *Of Acceptable Risk*. Los Altos: Kaufman.

Medford, D. (1973) *Environmental Harassment or Technology Assessment?* Amsterdam: Elsevier.

Mishan, E. J. (1977) Economic Criteria for Intergenerational Comparisons. *Futures*, 9 (5), pp. 383–403.

Office of Technology Assessment (1978) *Annual Report to the Congress for 1977*, Washington DC: OTA.

O'Riordan, T. (1976) Beyond Environmental Impact Assessment, in (eds.) O'Riordan, T. and Hey, R. D. (eds.) Chapter 15. *Environmental Impact Assessment*. Farnborough: Saxon House.

Rasmussen, N. C. *et al*. (1974) *Reactor Safety Study: An Assessment of Accident Risks in U.S. Commercial Nuclear Power Plants*. Main Report of the Nuclear Regulatory Commission. Washington DC.

Rose, R. (ed.) (1974) *Lessons from America*. London: Macmillan.

Routley, R. and V. (1978) Nuclear energy and obligations to the future. *Inquiry*, 21 (2), pp. 133–79.

Royal Commission on Environmental Pollution (1976) *Sixth Report: Nuclear Power and the Environment*, Cmnd. 6618. London: HMSO.

Sjöberg, L. (1979) Strength of belief and risks. *Göteborg Psychological Reports*, 7 (2).

Solesbury, W. (1976) The environmental agenda. *Public Administration*, 54 pp. 379–97.

Starr, C., Rudman, R. and Whipple, C. (1976) Philosophical basis for risk analysis. *Annual Review of Energy*, 1, pp. 629–62.

Statens Kärnkrafts Inspektion (1978) *Haveristudi av Barsebäck II*, SM/78/1 (Swedish Nuclear Power Inspectorate).

Stoel, T. B. and Scherr, J. S. (1978) Experience with EIA in the United States. *Built Environment*, 4 (2), pp. 94–100.

Swedish Energy Commission (1978) *Risk Assessment* (in Swedish), 1/1978: 15.

Town and Country Planning Association (1978) *Planning and Plutonium*, Evidence of the Town and Country Planning Association to the Public Enquiry into an Oxide Reprocessing Plant at Windscale. London: TCPA.

Tribe, L. H. (1973) Technology assessment and the fourth discontinuity: the limits of instrumental rationality. *Southern California Law Review*, 46, pp. 617–61.

Tribe, L. H., Schelling, C. S. and Voss, J. (eds.) (1976) *When Values Conflict*, Cambridge, Mass: Ballinger, especially Chapter 4, Frankel, C., The Rights of Nature.

Vlachos, E., Buckley, W. and Filstead, W. J. (1975) *Social Impact Assessment: an Overview*. Fort Collins: Colorado State University.

Waddington, C. H. (1977) *Tools for Thought*. St. Albans: Paladin.

Walker, C. A. and Black, J. S. (1978) The compleat assessor. *Social Science Energy Review*, 1 (4) pp. 26–52 (Yale University, Institution for Social and Policy Studies).

Walters, C. J. (1975) *Foreclosure of Options in Sequential Resource Development Decisions*. Laxenburg, Austria: International Institute for Applied Systems Analysis.

Wandesford-Smith, G. (1979) *Varieties of Environmental Impact Assessment: An International Analysis*. Berlin: International Institute for Environment and Society.

Watt Committee on Energy (1979) *Energy Development and Land in the United Kingdom*. London: HMSO.

Weinberg, A. M. (1972) Science and trans-science. *Minerva*, 10, pp. 209–22.

Index